Ethnomedicinal Plants
of Maring Tribe, Manipur

The Author

Dr Cheithou Charles Yuhlung completed his Graduation (2000) from St. Edmund's College, Shillong, M.A (2002) and was awarded Ph.D degree (2011) from the Dept. of Sociology, North Eastern Hill University (NEHU), Shillong, Meghalaya. Completed Post-Doctoral Research (2018) from the Dept. of Anthropology, Guahati University, Guwahati, Assam. He has worked as Research Assistant (2006) in an evaluation team for an International NGO (IWGIA) and as Research Associate (2011-12) in an UPE-Project, Dept. of Philosophy, NEHU, Shillong. He has served as Assistant Professor (Sociology) 2012-14 in Women's College, Shillong and also taught Sociology in William Carey University, Shillong, Meghalaya. He has published several original research papers in reputed journals like *Sociological Bulletin* (2007), *IOSR-JHSS, Tribal Bulletin, Asian Man and in others*. He served as the General Secretary of Northeast India Tribal Research Association (NEITRA) and had edited a book entitled as 'Northeast India Tribal Studies: An Insider's View'. He has published a book entitled as *'Chothe Indigenous Religions and Culture: A Sociological Study'* 2018. He has participated in various International and National seminars and conferences.

Ethnomedicinal Plants
of Maring Tribe, Manipur

(Published in completion of Post Doctoral higher research work under the UGC, New Delhi, scheme for SC/ST)

Cheithou Charles Yuhlung

2019
Regency Publications
A Division of
Astral International Pvt. Ltd.
New Delhi – 110 002

ISBN: 9789390384181 (HB)

Publisher's Note:

Every possible effort has been made to ensure that the information contained in this book is accurate at the time of going to press, and the publisher and author cannot accept responsibility for any errors or omissions, however caused. No responsibility for loss or damage occasioned to any person acting, or refraining from action, as a result of the material in this publication can be accepted by the editor, the publisher or the author. The Publisher is not associated with any product or vendor mentioned in the book. The contents of this work are intended to further general scientific research, understanding and discussion only. Readers should consult with a specialist where appropriate.

Every effort has been made to trace the owners of copyright material used in this book, if any. The author and the publisher will be grateful for any omission brought to their notice for acknowledgement in the future editions of the book.

Published by : **Regency Publications**
A Division of
Astral International Pvt. Ltd.
– ISO 9001:2015 Certified Company –
4736/23, Ansari Road, Darya Ganj
New Delhi-110 002
Ph. 011-43549197, 23278134
E-mail: info@astralint.com
Website: www.astralint.com

Digitally Printed at : **Replika Press Pvt. Ltd.**

Foreword

With the gradual declined in the practices of ethnomedicines among the indigenous tribal communities of the world Dr. Cheithou Charles Yuhlung has made his best effort in collecting the valuable information from the Maring people who are living in remote hill areas of the state of Manipur. He has scientifically and systematically organised about 60 identified plant species and a number of plants yet to be identified. The book provides a good glimpse on how and what kind of ethno-botanical plants were traditionally used by the Maring (Khoibu) tribe when required and as demanded. He has not only identified plants but also provided the treatment methods for various diseases by different parts of the plants. He has also highlighted how compound plants were used in certain cases to be more effective in the treatment procedure. In short, this is an interesting study on the practices of ethnomedicinal plants among the Maring tribe of Manipur. This exploration is just a beginning and it is hoped and expected that more of such applied research work will be done by the author in the near future. This book entitled as "Ethno-medicinal Plants of the Maring Tribe of Manipur, Northeast India", I am sure will trigger scientific interest and more researchers to follow the path of Dr. Cheithou Charles Yuhlung to document the vast traditional knowledge among the Maring and other tribal communities of Manipur before such knowledge vanishes completely. This interesting book will serve a good reference for students of ethno-botany and other research scholars and even laymen interested in the study of culture and ecology of the community.

I wish him success and best in life.

Prof. Mini Bhattacharyya Thakur
(Supervisor and former HOD)
Dept. of Anthropology
Guahati University,
Guwahati, Assam,
India - 780014

Preface

The book entitled "Ethnomedicinal Plants of Maring Tribe, Manipur" is based on the outcome of research work taken among the indigenous Maring tribe of Manipur. The universe of the study covers eleven villages located in three districts Viz: Chandel, Tengnoupal and Senapati. The data was collected using questionnaires, interviews and interview-scheduled based on a random sampling technique among select village elders, priest, priestess, village leaders and senior citizens of Maring by participant observation method.

The book is divided into five chapters. The Chapter One introduces the significance of the book. Chapter Two describes the identified indigenous medicinal plants of Maring tribe. Chapter Three describes the non-identified ethnomedicinal plants. Chapter Four is the analysis of the data presented in tables forms, while Chapter Five is the result and discussion that form the conclusion part.

In the study, there are a total of 95 (Ninety five) ethno-botanical plants, in which, 60 species are identified belonging to 37 families, while the other 35 plants are yet to be identified. The Asteraceae, Zingerberaceae, and Poaceae are the most common plant's families used by the Maring tribe with highest frequencies. This is because of the availability and the variety found in the region and for its remedy effectiveness. The statistics shows that the Maring suffered from 53 common sicknesses to certain vulnerable new diseases like fever, cough, dysentery to diabetes, jaundice, cancer, urinary/kidney stone cases, irregular menstrual problems, cardiac, liver and chronic ulcer, besides others ailments.

However, from the lists of Tables 4, 7 & 10 indicates that the Maring used more leaves of herbal plants which are seasonal in nature in the traditional form of decoction. The book also describes certain other forms of ethnomedicines like insects, animals, rice, and other important items in healing and curing certain ailments. It also highlights how the local medicine man/ woman uses compound plants items along with other substances in the treatment of different sicknesses.

Despite the advancement in pharmaceutical medicines the Maring still used their ethno-medicines side by sides in different ways to cure and heal certain ailments and diseases. There is lack of herbarium knowledge since; most ethnobotanical plants are available freely as grown wild on the roadsides and in the deep forest.

Some highly valuable seasonal plants are dried and kept safely at home for future used. Sadly, the practice of ethnomedicines among the Maring have considerably declined since most of the knowledgeable person have expired and the younger generation are not serious enough to learn or carry on the traditional knowledge of healing and curing methods. Certain steps need to be taken up to encourage the people to revive and continue their indigenous ethnomedicinal practices in a scientific manner. The study is basically a survey to understand how seriously and effectively the Marings practices their ethnomedicines in the contempory world.

Author

Acknowledgement

Firstly, I thanked the Almighty Heavenly God for all his blessings bestowed upon me in this endeavour in completion and protecting me during the course of my higher research work. I also expressed my gratitude to my family members and wife Elizabeth for their constant prayer, support and encouragement.

I am deeply honoured to Prof. Mini Bhattacharyya Thakur (former HOD), Dept. of Anthropology, Guahati University, Assam who happily consented to be my mentor and supervisor when I approached to her. Since then, she has been providing me invaluable suggestions and guidance, and she was always there for me whenever I needed her help. I also extend my sincere gratitude to Prof. Rekha Das (Retd.), and the then A.A Ashraf (HOD), all the faculty members, office staffs and fellow research scholars of the Department of Anthropology, Guahati University, Assam for their love, support and cooperation in the department.

The author sincerely expressed gratitude to the *University Grant Commission* (UGC), New Delhi for awarding him the Post-Doctorial Fellowship under the scheme of *Rajiv Gandhi Fellowship for the Scheduled Caste and Tribe* (SC/ST), awarded in 2012 for providing the financial assistance to do this higher research.

The author also felt indebted and expressed his sincere gratitude to all his Maring respondents like the village chiefs, elders, senior citizens, local medicine man, priest, priestess and leaders. This research work would not have been successfully completed without their kind cooperation and valuable information provided herein. To name some of the prominent respondents are like: (1) TL. Monal (67/M) S/o Kodang of Tuinem village, (2) Tangtanga Moshilpha (68/M) of Langol Khunjou, (3) Warok Menai (56/M) of Langol Khunjou, (4) Yunglama Thankang Muba (67/M) of Langol Khunou, (5) Ch. Medun (82/M) of Kharou Khunou, (6) Meirung Mejouba (78/M) of Khudei Khunou, (7) SK. Andun (68/M) of Khudei Khunou, (8) K. Tomui (78/F) of Kharou Khunou, (9) Saka Hoinu (65/F) of Kakching Lamkhai, (10) Saka Ringnga (33/F) daughter of S. Hoinu, (11) K. Tonmui (46/F) of Tuishimi, (12) Kansam Tungmui (W/o K. Medar) of Karamkhu village, (13) M. Meilung Mejouba of Khuringmul, and (14) M. Medun (56/M) of Sandang Senba.

The author also expressed thanked you to Miss. Chan Dawansha of Tuishimi, Mr. Sailung Mopham (Aboi) Maring of Khudei Khunou, Mr. Y. Stephen Damsu Chothe of Phaihu (Purum Tampak) and L. Morre Tarao of Leishokching for their assistances, guidance and introducing me to the Maring villagers for the survey. Besides, I also extent my thanked you to Dr. Lokho Abba Mao of Sankenikitan University, West Bengal for identification of ethno-botanical plants. Mr. N. Bijoy Singh, Mr. RK. Motilal and Mrs RK. Ranjita of Rajbari, Palace Compound, Imphal for helping me identify certain ethnobotanical plants in Manipuri (Meitei). I also expressed my gratitude to Mr. Dl Robinson of Lamlong Khunou and Dl. Angthil of Waksu, Tengnoupal District for helping me in verification of the plants.

I also expressed my sincere appreciation and thanked to certain friends like Mr. Heshu Aji Mao, John Kayina Mao, Dr. Jerry, Saleo, Jytishman, Arun, Nancy, Humi Thaosen, Chintu, Shangpam, and other friends who are always there for me and kept company during my higher research work and stayed in Guwahati, Assam. Lastly, but not the least, I expressed my sincere thanked you to all those persons whose names are not mentioned herein but have helped me in many different ways during the course of my work. Once again, in a special way, I thanked each and every one of you involved in this work and has had helped me in completion of this higher research work.

Thank You and God Bless You All

Yours Sincerely

Dr. Yuhlung Cheithou Charles

(Post-Doctoral Fellow)

Dept. of Anthropology,

Guahati University,

Guwahati, Assam – 780014

Abbreviations

Herb	– H
Shrub	– S
Tree	– T
Creeper	– C
Climber	– Cl
Seasonal	– S
Perennial	– P
Oral	– O
External	– E

Contents

Foreword *v*
Preface *vii*
Acknowledgement *ix*
Abbreviations *xi*
List of Tables *xv*

Chapter - 1
Introduction 1

Chapter - 2
Indigenous Medicinal Plants of Maring Tribe 5

Chapter - 3
The Practice of Ethnomedicines among the Maring Tribe 41

Chapter - 4
Analysis of the Data 59

Chapter - 5
Conclusion 89

Suggestions 99

List of Tables

Table 1:	Indigenous Medicinal Plants of Maring (Khoibu) Tribe	61
Table 2:	Distribution of Plant's Family	73
Table 3:	Distribution of Different Types of Ailments and Diseases from the Identified Ethno-botanical Plants	73
Table 4:	Distribution of Frequency of Identified Plant's Forms, Types, Parts Used and Forms of Usages	74
Table 5:	The Ethnomedicinal Plants of the Maring Tribe	75
Table 6:	Distribution of Frequency of Different Types of Ailments and Diseases of Maring	81
Table 7:	Frequency Distribution of Ethnomedicinal Plant's Forms, Types, Parts Used and Forms of Usages	82
Table 8:	Category Wise Distribution of Total Frequency of Ethno-botanical Plants of Maring	82
Table 9:	Distributions of Total Frequency of Common Types of Ailments and Diseases among Maring (and Khoibu) Tribe of Manipur	83
Table 10:	Category Wise Distribution of Total Frequency of Plant's Forms, Types, Parts Used and Usages	85
Table 11:	Other Forms of Indigenous Medicines Used by Maring Tribe	86
Table 12:	Preparation Methods of Some Compound Ethnomedicines Used by Maring Tribe	88

CHAPTER - 1

◆◆◆◆◆

Introduction

The practice of ethno-medicines has been since the beginning of human society. It is widely practice by different communities and tribes of the world in different forms. However, with the development in science and technology, advance in pharmaceutical medicines, influences of modernization, westernization, proselytization to Christianity, disturbance in environments, changed in climate, and gradual declined in the number of traditionally knowledgeable individuals have resulted in the decline of the practice of traditional ethno-medicine. Because of such factors of inevitable global developments and changes it has posed a serious threat to the rich traditional knowledge of different ethnic groups and tribes of the world. Therefore, a survey was focus in documenting and understanding the extensiveness and its relevancies in the use of various ethno-botanical plants as medicines among the Maring tribe of Manipur.

Brief History of Maring

The Maring (and Khoibu) is one of an indigenous tribal group of Manipur. They are predominantly located in Chandel, Tengnoupal and Senapati Districts of Manipur. Khoibu comes under the sub-group of Maring. The internal distinction is because of certain internal variation in term of languages, dress codes, customs and traditions, which is common among other tribal groups. However, for the study we shall refer to as Maring only since it is listed in the Scheduled Tribe list of Government of India, while the latter is not yet recognised.

The term 'Maring' (Murrings) is derived from the words '*Mei*' means fire, and '*Ring*' means to start or produce, or sometimes termed as '*Meiringba*' which means 'The people who keep the Fire unquenched'.[1] Linguistically, Dr. Grierson in his book *Linguistic Survey of India* has classified the Maring language/ dialect under the sub-group of Naga-Kuki languages under the category of *Tibeto-Burman* linguistic family.[2] The Census of India (COI) 2011 gives the Maring population as 26,408 with a literacy rate of 55.12%.[3] With the advent of Christianity many have converted to Christians. Culturally, they are distinguishable from other cultural groups because of their peculiar tradition of blacken teeth, knotted-hairs (*Pagri*/ Turban/ *Coiffure*) tied with white and wrapping saffron cloth on the forehead side and wearing their traditional white and black attires with heavy big ear-rings.[4]They are considered one of the closest cognates or relatives of Meitei (Manipuri) since they claimed to inhabit in and around the Imphal (Kangla) kingdom in the early days. They are known for their bravery, warrior attitudes and practice of deadly magical witchcrafts against their enemies in the past.

However, with the advancement in modernisation, westernisation, globalisation and the impact of science and technology the majority of Maring (and Khoibu) tribe have changed their outlook, attitudes, culture and traditions to suit the modern world from traditionalism. They are fast transforming and the transition is shaping up their society from traditionalism to modernity at high rate.

Literature Review

The North-eastern region of India comprises eight states; Assam, Arunachal Pradesh, Nagaland, Manipur, Meghalaya, Mizoram, Tripura and Sikkim. The region comes under the lower Himalayan hill ranges and is known for its remarkable biodiversity. The region is located in temperate tropical rain forests zone within 23°51′ N and 25°41′ N Latitudes, and 93°3′ E and 94°4′ E longitude bordering Myanmar to the east, China and Nepal on the north.[5] The region is richly supported with diverse flora, fauna and several crop species. The region is ranked 8th amongst the 34 'Bio-diversity Hotspots' in the world.[6] The region is not just rich in biodiversity but also very rich in 'cultural diversity' because there are about 175 distinct languages (dialects) spoken indicating of different cultural groups. But if sub-groups are included then, there are more than 200 tribes and communities inhabiting in the region that uses different ethno-botanical plants as medicines based on their belief and practices in curing and healing of various ailments and diseases.[7] Despite the advancement in modern pharmaceutical medicines many rural and urban people still depends upon some common traditional ethno-botanical plants as medicines for its effectiveness in certain cases of sicknesses, ailments and diseases.

Although, of lately there has been tremendous encouragement to survey and de-code from the Government and other agencies on the practice of ethno-medicine on different communities of the world for its relevancies in its application. However, to scientifically identify the main constituent elements and components of the identified ethno-botanical plants will take several years more for the successful remedy and curative effect on certain common ailments and deadly diseases.

A.A Mao (2009) said that since 1970's different universities and research institutions began to study ethnobotany and traditional knowledge system and later by 1980's the *Ministry of Environment and Forests*, Govt. of India launched a project to document ethno-botanical uses of plants by indigenous people of India. Under the project *Botanical Survey of India,* Eastern Circle, Shillong undertook to study all the North-eastern states in phase manner.[8] The continued project is expanding in other states of Northeast India.

Since, there has been exponential growth of interest in the treatment against different diseases using herbal drugs as they are generally non-toxic and *World Health Organization* (WHO) has recommended the evaluation of effectiveness of plants in condition where we lack modern safe drugs.[9] Rajkumari (2013) in her study among the Chiru tribe said that due to poor transportation and non-available medical facilities they still depend on the traditional faith upon the local medicine men and wild herbal plants for survival.[10] With such factual information the WHO estimates that over 80 per cent of people in developing countries depend on traditional herbal medicines for their primary health needs.[11] There are as many as 1,200 plants identified and used as medicine in ancient Indian texts.[12]

Now, ethnobotany and ethno-medicinal plant studies are well recognized as the most viable method for identifying new medicinal plants and refocusing on those earlier reported for bioactive constituents.[13] For example, getting direct health benefit from eating fresh ethno-botanical plants is encouraged by Yumnam and Tripathy (2012). They said that the Meitei (Manipuri) tradition of eating parts of raw plants mixed with other edible ingredients called *Singju,* is believed to have direct medicinal benefits on the person.[14] The North-eastern region is an important part of the Indian Floristic Zone and has been identified as one of the twelve 'Genetic Epicenters' for the evolution of world flora.[15] Therefore, there are over 500 species of medicinal plants reported from Arunachal Pradesh and an estimate of around 8000 manufacturing units of traditional medicinal systems in India.[16]

However, with regards to the rareness and extinction of ethno-botanical plants with medicinal values Jamir (2012) reveals that many of these valuable plants are under threat and depleting very fast owing to rampant destruction of forests, practice of 'Jhum' or 'Shifting' cultivation, forest fire, over-exploitation of plant resources and other human socio-economic developmental activities in the region.[17] This applies not only to Nagaland state but to all states of North-eastern region especially, the unexplored small tribes located in the deep interior hilly regions like Maring, Chothe, Tarao, Aimol, Karam, Moyon, Monshang, Chiru, Koireng, Liangmei, Mao, Maram, Poumei, Tangkhul, etc. who mostly listed as endangered tribes that need immediate attention to document and assess the status of their wild edible and ethno-medicinal practices at the earliest. The urgency is that the village elders and local medicine men who are the knowledge bank and technical know-how persons are decreasing every year.

In this context, Jamir also suggested an urgent need for conservation and protection of biodiversity including the precious wealth of medicinal plants in the region with a view that detailed investigation by modern scientific techniques, at least some of these plants might prove to be effective life-saving drugs plants for cure and relief of various ailments and diseases that plague to mankind.[18] Further, Neli Lokho Pfoze (2012) in his study among the Mao people of Senapati district found that there has been no report of cultivation of medicinal plants by the local peoples and the preparations are made by collecting the plants from the wild. He said this is a serious concern from the point of conservation and sustainability of the resources because such collection from the wild may lead to depletion of the plant population or even extinction of the resources particularly the rare and endangered species if it goes unabated.[19] Similarly, A.A Mao asserted that most of the wild vegetables, fruits and medicinal plants uses are of little known or not known at all to the outside world. Also, many of the known medicinal plant in used have not been studied empirically in detailed for the active chemical compounds.

Therefore, he laments that despite India being second to none in documentation in this field of ethno-botanical research publication, the country has so far not been able to translate the information into economic wealth.[20] Such statements need deep introspection by the concern authorities in the related fields. So far, only the 'Patanjali Ayurved' Ltd. founded by Baba Ramdev is running success besides few others, but where are the other hundreds and thousands expertise gone? How do we trap such ethno-botanical plants for effective use to save millions of lives is the question hour now?

Objectives

The basic objectives of the study were to survey and understand the use of ethno-botanical plants used by the Maring (and Khoibu) tribe as medicines, since no record is available with regards to the subject matter so far. It was also to find the intensive and extensiveness of the used these ethno-medicines among them in curing and healing various ailments and sicknesses. The practice of ethno-medicine is declining very fast due to various global factors of changes and increase number of deceased of knowledgeable elderly persons every year.

Materials and Methods

The field survey was carried out among the 11 select Maring villages located in Chandel, Tengnoupal and Senapati districts of Manipur. The villages are: 1) Kharou Khunou and 2) Langol Khunjou located in the deep interior bordering Myanmar which is about 130 Kms from Imphal. 3) Langol Khunjou, 4) Khunbi and, 5) Langol Khunou are situated on the top of the eastern hill ranges which is 500 meters above sea level. 6) Maring Sandang Senba of Senapati district is situated on the northeast hilltop overlooking Andro village. 7) Khudei Khunou, 8) Tuinem, 9) Tuishimi, (10) Khuringmul, and 11) Kakching Lamkhai of Khoibu villages are located on the foothills in and around Pallel near the valley.

Initially, formal request was made to the respective village chiefs and leaders by introducing ourselves and explaining the purpose of our visit to them. Thereafter, interviews and interview scheduled were conducted in phase manner individually and in groups among the 15 selected respondents (10 males and 5 females). Although there were more respondents in reality but due to their vague information they have been neglected in the discourse. The respondents are like village priests, priestess, village elders and senior citizens who are knowledgeable in the practices of ethno-medicines. The researcher also took some photographs of the identified plants during the fieldworks.

Secondary sources of data like published articles, papers, books, websites are also referred for the data analysis, especially in identifying the scientific names of the ethno-botanical plants collected.

CHAPTER - 2

◆◆◆◆◆

Indigenous Medicinal Plants of Maring Tribe

1. Botanical Name: *Adhatoda vasica* Nees. [**Family**: Acanthaceae], **Common Name**: Malabar nut. **Local Names - Maring:** *Triptung-ngou/ Flupbu/ Simlim khangou paar.* **Manipuri:** *Nongmangkha Angouba.*

Diseases/Treatment: (a) *Fever,* (b) *Cough.*

Parts used: Leaf.

a) Crush the leaf and collect the liquid and add little honey and serve a spoonful twice a day.

b) Crush leaves are mixed with kerosene and apply at the back of the earlobe or boil the leaves and may also bath with it to relief from cough.

How to identify: This shrub *Adhatoda vasica* is another species of Acanthaceae family, which is very similar to Phlogacanthus thyrsiflorus Nees or Basak plant because they have similar taste, while this has white flower and the other yellow flower.

2. Botanical Name: *Aeschynanthus hookeri* C.B.Clarke, [**Family**: Gesneriaceae], **Common Name: Local Names - Maring:** *Anlikli.* **Manipuri:** *Utangbi.* **Diseases/ Treatment**: Diabetes.

Parts used: Leaf, bark.

A handful of this *Utangbi* leaves are boil and served twice as decoction one glass in empty stomach before each meal till the diabetes is cured.

How to identify: This plant is not exactly *Aeschynanthus hookeri* species of the family Gesneriaceae since there are about 150 species in the family. But like any *Gesneriaceae* plants species this plant also gets nutrition from another tree by hooking it. Therefore, it is locally known as *Utangbi* literally means 'hooky plant that gets support from a tree' in Manipuri. There are other species of *Gesneriaceae* plants which are not recommended for any treatment. This plant has small white and yellowish flowers at each leaf knots near the fiberous hanging roots. The leaves are hard and smooth. The stems are also hard. The ants used to gathered around the tree branch where it grew bushy.

3. **Botanical Name:** *Agaricus campestris* [**Family**:], **Common Name**: Meadow/Ground mushroom, **Local Names - Maring:** *Thrai-meetlung/ Leibak-marum,* **Manipuri:** *Leibak-marum.* **Diseases/Treatment**: Burn.

Parts used: Whole plant.

Immediately apply the extracted liquid of the ground mushroom or the powder obtained on the burned part softly.

How to identify: This algae grows where there was some humus left on the dry soil and the area becomes wet occasionally. In certain cases it is found grown on the meadows.

4. Botanical Name: *Ageratum conyzoides* Linn. [**Family**: Asteraceae], **Common Name**: Goat weed. **Local Names - Maring:** *Yanglou*. **Manipuri:** *Lounamba/ Khongjai napi*

Diseases/Treatment: Cuts, wounds.

Parts used: Leaf.

Crush some fresh leaves and apply it on the cuts or wound areas.

How to identify: This goat weed grows wild on roadsides or in open wasteland areas.

a)

b)

5. Botanical Name: *Albizia myriophylla* Benth. [**Family**: Leguminosae], **Common Name**: Little-leaf sensitive-briars. **Local Names - Maring:** *Threlou/ (a) Silip yangyu, (b) Rui yangyu.* **Manipuri:** *Yanglee.*

Diseases/Treatment: (a) Stone Case/ Kidney, (b) Dog Bite.

Parts used: Bark.

a) Boil some leaves of *Albizia myriophylla (Yanglee)* with *sitamasi* (white sugar free) and have three spoonful four times a day. The urine will be black in colour.

b) Chew the bark of *Albizia myriophylla* with little raw rice and apply the paste immediately at the dog's bitten part.

How to identify: The *Albizia myriophylla* species is of two varieties: a) one is a shrub plant and the other is b) a climber plant. The plants bear small whitish flowers and the fruit is like small tree beans. The bark taste little sweet to gummy and gives giddiness. The bark has little ferment smell. It grows in deep forest and seldom found now. The dry bark of the plant is used in preparation of local yeast for brewing local wines and rice-beers.

6. Scientific Name: *Alocasia macrorrhiza* (Linn.) Schott. [**Family**: Araceae], **Common Name**: Giant taro. **Local Names - Maring:** *Madung/ Lam pan.* **Manipuri:** *Hongngo.*

Diseases/Treatment: Bee sting.

Parts used: Stem.

Cut the stem and immediately apply the extracted liquid on the bee stung area. One should be very careful not to touch the extracted liquid with hand for it is very itchy.

How to identify: It is herb plant wildly grown in the wasteland areas on the roadsides or swampy gardens. The leaves are huge and shiny. It bears pink fruit which is uneatable. This wild yam is not eatable.

7. Botanical Name: *Alpinia allughas* Roscoe. [**Family:** Zingiberaceae], **Common Name**: Galangal. **Local Names - Maring:** *Puleimal.* **Manipuri:** *Puleimanbi.*

Diseases/Treatment: a) Gas Formation (Flatulence), b) Ascaris worms in children. **Parts used:** Rhizome.

a) Crush the roots (rhizome) of *Alpinia allughas* and mix with some crush gooseberry liquid and add little honey and take one tea spoonful as syrup after food whenever thirsty. It helps reduced gas formation and digestion as it act anti-oxidant agent.

b) Extracted liquid is mix with little tobacco or kerosene and apply all over the body and the lower throat to dispel the intestinal worm especially for children and adults too.

How to identify: The *Alpinia allughas* Roscoe though has similar leaf structure is thinner and smoother compared to other Zingiberaceae species. The flower blooms in bunch of elongated shape mixed in pink, yellow and white colours. If crush the leaf it will give you a strong spicy aromatic smell like the root that grew wide spread like a stem with many knots in the soil. Little amount of the fresh/dry roots are added in fish and pork curries to give strong spicy smell.

8. Botanical Name: *Alpinia galangal* Linn. Willd. [**Family:** Zingiberaceae], **Common Name**: Greater galangal.**Local Names- Maring:** *Ramrhou.* **Manipuri**: *Kanghu.*

Diseases/Treatment: Piles.

Parts used: Rhizome.

Crushed small amount of *Alpinia galangal* along with some tobacco leaves and is then inserted inside one's anus. Do it twice a day till cured. (Some other communities suggest to drink the crushed liquid in an empty stomach, or eat the dry rhizome in powdered form every day till cured).

How to identify: This *Alpinia galangal* species though has similar leaf structure with other *Zingiberaceas* species. It can be differentiated by the small reddish and white flower bloom in bunch bearing small bulb kind of seeds. When crushed the leaf smell lesser spicy aromas.

9. Botanical Name: *Antidesma acidum* Retz. [**Family**: Euphorbiaceae/ phyllanthaceae], **Common Name**: Amita.**Local Names - Maring:** *Antuitik/ Ram Ansur.* **Manipuri:** *Ching-yensil.*

Diseases/Treatment: Arthritis/ Rheumatism (Joints pain).

Parts used: Leaf.

Boil the leaves of *Oxalis corniculata* and served as decoction a glass full three times a day till cure. It may also be served as culinary item. (Fruit is black in colour when ripen).

How to identify: The *Oxalis corniculata* is a shrub plant grown especially under the shade of a large trees in the deep tropical forest area where humid is high. The leaf has acidic or sour taste like that of yellow Wood Sorrel (*Oxalis stricta*). The hill tribes of Manipur eat the leaves as curry with dry fish and meats apart from medicial uses.

10. Scientific Name: *Areca catechu* L. [**Family**: Palmaceae], **Common Name**: Areca/ Betel nut. **Maring Name:** *Kom kwa.***Manipuri Name:** Kom kwa.

Diseases/Treatment: Menstrual Problem.

Parts used: Seed/ Nuts.

Spelt upon the beetle nut and cut into half and throw away. (This involves use of magical charm while breaking).

11. Botanical Name: *Azadiracta indica* A. Juss [**Family**: Meliaceae], **Common Name**: Neem plant.**Local Names - Maring:** *Neem-trung.* **Manipuri:** *Neem.*

Diseases/Treatment: Malaria.

Parts used: Leaf.

The extracted liquid obtained by crushing the leaves are mix with little water and about 2-3 tea spoonful is taken twice a day before each meal. (Do not give to those who are weak/ unhealthy/ pregnant or those suffering from leukaemia since it is extremely bitter).

12. Botanical Name: *Benincasa hispida* Thunb.Cogn. [**Family**: Cucurbitaceae], **Common Name**: Ash-gourd/ Winter melon. **Local Names - Maring:** *Anmahei-angou /Kulbi.* **Manipuri:** *Torbot.*

Diseases/Treatment: Bear/ Tiger Bites.

Parts used: Fruit.

After peeling the covering of the *Benincasa hispida* immediately apply the paste on the bitten area and lightly bandage it. One may also sprinkle the powdered *Kursi* seeds upon the wounded area for effectiveness.

How to identify: The leaves of this ash-gourd creeper plant are little less rough and spiky than pumpkin plant. The outer fruit covered with white ash type and the inside is colourless and cold most of the time. The hill variety is considered much better in taste than those grown in the valley.

13. Botanical Name: *Blumea balsamifera* D.C. [**Family**: Asteraceae], **Common Name**: Elumea or Nagal Camphor. **Local Names - Maring:** *Langthrei.* **Manipuri:** *Langthrei.*

Diseases/Treatment: Burning Sensation of Stomach.

Parts used: Leaf.

Crush some amount of the leaves into paste form and mix with little water in a glass and drink it immediately anytime till it cures. Or one may chew few leaves and drink water to get relief immediately.

14. **Botanical Name:** *Carica papaya* L. [**Family**: Caricaceae], **Common Name**: Papaya.
Local Names - Maring: *Tengshangbo/ Khnachang hei.*
Manipuri: *Awathabi.*
Diseases/Treatment: Ulcer.
Parts used: Fruit.

Boil the unripe papaya and have regularly as culinary or anytime. One may also have the ripe papaya regularly anytime till it helps relief.

15. Botanical Name: *Centella asiatica* Linn. [**Family**: Apiaceae], **Common Name**: Indian pennywort.
Local Names - Maring: *Anlaiphun.* **Manipuri:** *Peruk.*
Diseases/Treatment: Sore Throat / Hypertension.
Parts used: Whole plant.

Boil certain amount of the whole plant *Centilla asiatica* in about two litres of water and have as decoction a glassful (200 ml) three times a day. Repeat till it is cured. One may also have the fresh leaves regularly as culinary item to reduce hypertension or high blood pressure. This creeper serves as anti-oxidant and served as valuable in the treatment of other ailments too.

16. Botanical Name: *Chenopodium ambrosioides* Linn. [**Family**: Chenopodiaceae], **Common Name**: *Wormseed* / Jesuit's tea / Mexican tree. **Local Names - Maring:** *Kerosene-numbi.*

Manipuri: *Kerosene-numbi/ Tersing numbi.*

Diseases/Treatment: Worm for babies.

Parts used: Leaf.

Crush the leaves and apply on the baby's body as balm this will repel the intestinal worms, especially around the chest. Repeat for three-four days.

How to identify: This herbal species of *Chenopodium ambrosioides* may not be exactly the same with what is available out here in Manipur. This herbal plant leaves when crush has the petroleum or kerosene smell. Therefore, the local called it as *Kerosin-numbi* literally 'Kerosene smelly plant'. The herbal plant grows during the months of December-April. It bears small whitish and yellow small flower for only few days. It grows in fertile wasteland areas near human settlements. There are other similar species.

17. Botanical Name: *Cinnamomum camphora* (Linn.) Nees & Eberm. [**Family**: Lauraceae], **Common Name**: Camphor tree.

Local Names - Maring: *Kapot.* **Manipuri:** *Karpur pambi.*

Diseases/Treatment: Post-natal women, annemia.

Parts used: Leaf.

The leaves are boiled and taken bath by post-natal woman who had delivered baby. It helps relief body ache and from lack of annemea. This is done twice or thrice a week.

How to identify: The leaves of *Cinnamomum camphora* shrub plant are thick, soft and hairy with little awful smell. It grows in dry tropical upland areas where it gets sufficient sunlight.

18. Botanical Name: *Cinnamomum zeylanicum* Breyn. [**Family**: Lauraceae], **Common Name**: Ceylon cinnamon.

Local Names - Maring: *Ushingsha.* **Manipuri:** *Ushingsha.*

Diseases/Treatment: Cough, Tonsilitis.

Parts used: Bark.

Boil little amount of the dry *Cinnamomum zeylanicum* bark in about half a litre of water for sometime and add little Sitamasi (White sugar cube) and served warmly as decoction three times a day after food till relief.

19. Botanical Name: *Colocasia esculenta* (L.) Schott. [**Family**: Araceae], **Common Name**: Yam. **Local Names - Maring:** *Baal.* **Manipuri:** *Pan.*

Diseases/Treatment: Spike on Heel.

Parts used: Rhizome.

Cut the fresh yam and paste on it or bandage around the heel for two days and repeat the same process till heal.

How to identify: This yam is the common yam eaten by most people which is like tuber.

20. Botanical Name: *Curcuma caesia* Roxb. [**Family**: Zingiberaceae], **Common Name**: Black zedoary.**Local Names - Maring:** *Aidai/...* **Manipuri:** *Yaimu.*

Diseases/Treatment: (a) Vomiting of Blood, (b) Menstrual Cycle Problem, (c) Child Indigestion.

Parts used: Rhizome.

(a) For person vomiting with blood: Mix the crushed *Curcuma caesia* liquid with little water and have half a glass before each meal twice a day. The bad blood will be diffused immediately with stool. Repeat it till it helps cure. (b) Menstrual Cycle Problem: Crush certain amount of *Curcuma caesia (Yaimu)* and add the liquid with some water and take 3-4 tea spoonfuls twice a day for a week before each meal. (c) Child Indigestion: Collect the crushed liquid of some amount of *Curcuma caesia* and add some water and give half tea spoon for child below 3 years and 1 tea spoonful above 3 years old. One may also apply the crush paste on the child's body, especially on chest and back (not stomach). The child faeces will be very smelly.

How to identify: The plant grows in bunch with short stalks and bears only a only a small amount of rhizome that is dark grey/black in colour inside, therefore the locals called it as *Yaimu* (lit. black turmeric). It has not much aromatic smell but when tasted it gives a very concentrated kind of awful bitter taste like some medicines. The species is not exactly *Curcuma ceasia* but closer to it because the flowers are very small and bluish in colour. The leaves are like turmeric leaves and grows in clusters. Considered a kind of ginseng though different in plant structure.

21. Botanical Name: *Curcuma C. zedoaria* Roscoe. [**Family**: Zingiberaceae], **Common Name**: Wild white turmeric/ Zedoaria/ Theap Rum Ruk.**Local Names - Maring:** *Aitonsan angouba.* **Manipuri:** *Yai-ngang angouba.*

Diseases/Treatment: Pigmentation/ Black/ Dark Spot/ Pimples.

Parts used: Rhizome.

Crush the *Curcuma longa* and add little mustard oil and apply the paste regularly on the affected area before bedtime for a week.

How to identify: This white turmeric plant grows wild and is rarely found. The plants and leaves are larger than normal turmeric plant. The Rhizome is large and white in colour. This species may not be exactly the *curcuma australasica* but appears to be closer than the *zedoaria* species.

22. Botanical Name: *Curcuma longa* Linn. [**Family**: Zingiberaceae], **Common Name**: Turmeric. **Local Names - Maring:** *Aitonsan.* **Manipuri:** *Yai-ngang.*

Diseases/Treatment: Cuts/ Wounds.

Parts used: Rhizome.

Crush the fresh *Curcuma longa* (turmeric) and immediately apply on the cuts or wounds. It helps in blood coagulation and heals fast. It is considered one of the best antiseptic in early days.

23. Botanical Name: *Cymbopogon citratus* (DC.) Stapf. [**Family**: Poaceae], **Common Name**: Lemon grass.**Local Names - Maring:** *Lemon grass.* **Manipuri:** *Lemon grass.*

Diseases/Treatment: (a) Sinusitis, (b) Ringworm, (c) Cardiac and diabetic.

Parts used: Leaf.

Crush the leaves and apply the *Cymbopogon citratus* liquid around the nose. One may apply the liquid around the infected ringworm area. For cardiac and diabetic patients, boil some fresh or dry leaves and serve directly as green tea.

How to identify: This herbal Cymbopogon citrates plant is like a thatch plant but the leaves has strong aromatic sweet smell when crushed.

24. **Botanical Name:** *Cynodon dactylon* (L.) Pers. [**Family**: Poaceae], **Common Name**: Bermuda/ Durva grass. **Local Names - Maring:** *Phaiphong*. **Manipuri:** *Tingthou napi.*

Diseases/Treatment: a) Typhoid, b) Baby's Measles and white stool.

Parts used: Whole plant.

a) Crush well some amount of the tender leaves of *Cynodon dactylon*. For effectiveness the liquid collected in half a glass is mix along with some crush pomegranate leaves and add some honey and take three times a day before food twice a day. It is considered effective for children.

b) Crush the *Cynodon dactylon* and mix with little alcohol and apply on the whole baby's body repeatedly three four times a day till cured. It helps children suffering from measles and white stools.

How to identify: This grass especially grows on the lawn or courtyard. The species is usually a small variety of Durva grass where the small and smooth leaf grows out from each knots in opposite direction. When crushed the plant has a strong mud or sand small. The stem has the structure of sugarcane knots.

25. Botanical Name: *Desmodium canadense* (Linn.) DC. **[Family:** Fabaceae**], Common Name:** Showy/Canada tick-trefoil.

Local Names - Maring: *Kakching threilou.* **Manipuri:**

Diseases/Treatment: Menstrual problem.

Parts used: Leaf.

Boil the leave and serve as decoction a glassful twice a day after food till cured. (The leaves can be served as tea).

How to identify: This herbal plant is wildly grown in the dry tropical upland areas, especially on the mound. The leaves are coarse, rough and thin. The flowers are bluish to pinkish and the seeds have beans shape port. The *Desmodium canadense* is the nearest species to this plant.

26. Botanical Name: *Desmodium rotundifolium.* **[Family:** Fabaceae**], Common Name:** Ticktrefoil. **Local Names - Maring:** *Mualou.* **Manipuri:**

Diseases/Treatment: Blood purifier (esp. for Post-natal women). **Parts used**: Whole plant.

Boil and serve as decoction a glassful three times a day before food. One may also have a bath with the extracted boil water. It even helps those regular menstrual for girls.

How to identify: This herbal plant is not exactly the species of *Desmodium rotundifolium* although belongs to Fabaceae family. It grows in cold moisture areas in the deep forest and roadsides of the country sides. The flowers are small and white to bluish to light purple in colour. The leaves are soft, tender and oval in shaped. Average high of the plant is ½ -1 foot.

27. Botanical Name: *Eclipta alba* (Linn.) Hassk. [**Family**: Asteraceae], **Common Name**: Trailing eclipta. **Local Names - Maring:** *Uchi sumban.* **Manipuri:** *Uchi sumban.*

Diseases/Treatment: Fever, Cough.

Parts used: Whole plant.

Crush the leaves and the extracted liquid is mixed with little honey and given for cough and fever two tea spoonful twice a day.

How to identify: This *Eclipta alba* herbal plant is grown in the tropical temperate areas especially around the waste dry places where there are moistures. The leaves are rough with green buds bearing white small flowers.

28. Botanical Name: *Elsholtzia blanda* Benth. [**Family**: Lamiaceae], **Common Name**: Lengser Rep. **Local Names - Maring:** *Tumpina/ Teckta.* **Manipuri:** *Lomba.*

Diseases/Treatment: Boil.

Parts used: Leaf and flower.

Heat up the fresh leaves of *Elsholtzia blanda* and apply on the boil part with little opening at the boil mouth. Repeat till it is cured.

How to identify: The species *Elsholtzia blanda* has a slight Vicks aromatic smell. Besides, medicinal properties the herbal plant is served as a delicacy item in culinary especially, with yam and fish curry by many Northeast tribes.

29. Botanical Name: *Eupatorium odoratum* (L). [**Family**: Asteraceae], **Common Name**: Siam/ Devil weed.

Local Names - Maring: *Muitem/ Loukhamong.* **Manipuri:** *Chelmen napi*

Diseases/Treatment: Cuts and scratches.

Parts used: Leafs.

Boil and have as decoction three times a day a glassful after each food. Crush the leaves and apply on the cuts and scratches to stop blood oozing. It acts as antiseptic.

How to identify: It grows widely in open waste land areas. It has a little awful smell. It may not be the exact species of *odoratum* but closest to it.

30. Botanical Name: *Ficus auriculata* (Lour.) Roxb, [**Family**: Moraceae], **Common Name**: Fig tree. **Local Names - Maring:** *Khrut hei.* **Manipuri:** *Heirit.*

Diseases/Treatment: Ringworm.

Parts used: Leaf.

Scratch the ringworm area with the fig leaf slightly at an interval time till cured then apply the ringworm area with the young bamboo shoot peeled cover.

31. Botanical Name: *Garcinia xanthochymus* Hook. [**Family**: Clusiaceae], **Common Name**: Dampel (Hindi).

Local Names - Maring: *Channahei* (Big leaf variety). **Manipuri:** *Heibung-asinba.*

Diseases/Treatment: Body Swelling.

Parts used: Leaf.

Heat certain amount of the *Garcinia xanthochymus* leaf on the fire and use it as balm on the swelling area. It soothes and helps reduce the swelling. Even the fruit can be used by heating up as balm.

How to identify: The plant grows in deep forest zones. The fruit is extremely sour with thick covering like elephant skin. Therefore, some commonly called it as elephant fruit.

32. Botanical Name: *Gynura cusimba* (D. Don) Moore. [**Family**: Asteraceae], **Common Name**: Hill Gynura. **Local Names - Maring:** *Louran.* **Manipuri:** *Terapaibi.*

Diseases/Treatment: Hypertension, high blood pressure, anti-inflammation.

Parts used: Leaf.

One may have fresh leaves as chutney or boil the leaves as culinary and have regularly to reduce the high blood pressure and hypertension. It also acts as anti-inflammation of the stomach.

How to identify: This herbal plant grows wildly in open wasteland areas even around the road sides. The matured seeds with fibrous cotton like thread are blown spinning away when gushy wind comes.

33. Botanical Name: *Gynura procumbent* (Lour.) Merr. **[Family**: Asteraceae], **Common Name**: Subungai. **Local Names - Maring:** *Tonglaibo*. **Manipuri:** *Ukhajing.*

Diseases/Treatment: Diabetes.

Parts used: Leaf.

Boiled some amount of the tender leaves for about 15-20 minutes in about two litres of water and served as decoction thrice a day after each meal. The leaves can be eaten fresh as salad or along with chutney or as simple boil curry.

How to identify: This particular *Gynura procumbent* species is grown in clusters near the houses and forest areas. This herbal plant is deep green in colour, soft and fleshy with small white flowers that bloom for short time. The leaf is tender and slipery and smells like grasses. It can be cultivated with the stem. The plant is considered rich in iron and minerals. (This is not exactly Ukhajing plant what most Manipuri knew but considered one species of its kind said the informant).

34. Botanical Name: *Kalanchoe K. farinacea* Balf.f **[Family**: Crassulaceae], **Common Name**: Flaming Katy.**Local Names - Maring:** *Manna-houbi.* **Manipuri:** *Manna-houbi.*

Diseases/Treatment: High Fever.

Parts used: Leaf.

Crush the leaves and apply the paste on the forehead. Do not eat. This is for external use only especially for adults.

How to identify: This species belongs to the Crassulaceae family, although the exact species may be different from others. The flower is white and pinkish in colour and blooms in bunches. It has the cooling effect.

35. Botanical Name: *Lilium sp.* [**Family**: Liliaceae], **Common Name:Local Names - Maring:** *Thrunlou.* **Manipuri:** *Lin-napi* /.

Diseases/Treatment: Snake bite /Dog bite.

Parts used: Tuber/bulb.

Crush well this tuber (onion type) and apply on the bitten spot immediately. Do not mix with water or taste the crushed item. (During medication one should not take fruits. It is for external use only, since it is considered highly poisonous.

How to identify: This herbal *Thrunlou (Lin-napi)* is a wild evergreen plant grown in temperate upland areas among the grasses. The tuber is like an onion and the leaves are like small palm leaves and blooms like small white lily flowers for only a week. Some people also cultivated at home as garden flowers. Though belongs to the genus *Lilium* of the family Liliaceae the exact species cannot be ascertain now.

36. Botanical Name: *Mangifera indica* L. [**Family**: Anacardiaceae], **Common Name**: Mango. **Local Names - Maring:** *Nouwa hei.* **Manipuri:** *Heinou.*

Diseases/Treatment: Diarrhoea/ Dysentery.

Parts used: Leaf.

Boil some mango leaves with some water and drink half a glass 2-3 times a day after food. Especially, recommended for adults.

37. Botanical Name: *Melastoma malabathricum* Linn. **[Family**: Melastomataceae], **Common Name**: Matakui (Palau). **Local Names - Maring:** *Umba heitrung.* **Manipuri:** *Yachubi.*

Diseases/Treatment: Debetis:

Parts used: Leaf, Stem.

Boiled some leaves of *Ombouheitrung* in about two litres of water and served as decoction a glassfull twice a day before each meal till cured.

How to Identify: There so many varieties. The plant with the purple-pink flowers and the fruit are small bud like with small spikes is edible that stains.

38. Botanical Name: *Melothria maderaspatana* (L.) Cogn. [**Family**: Cucurbitaceae], **Common Name**: Wild cucumber. **Local Names - Maring:** *Ram machanghei/ Bemangjam.* **Manipuri:** *Lamthabi.*

Diseases/Treatment: Jaundice, Mal-nutrition.

Parts used: Whole plant.

Boil the whole fresh creeper plant of *Melothria maderaspatana* or the dry one in about two litres of water and serve as decoction a glass full before each meal once/ twice a day.

How to identify: This creeper plant is found grown only in the deep temperate forest. There are two common types of varieties used. The leaf and fruits are spiky. The small fruit is little sour and is eatable.

39. Botanical Name: *Musa paradisiaca* Linn. [**Family**: Musaceae], **Common Name***:* Banana flower. **Local Names - Maring:** *Mut-thro.* **Manipuri:** *Laphu tharo.*

Diseases/Treatment: Dysentery. **Parts used:** Flower (Young fruit).

Slightly roast the *Musa paradisiaca the* banana fruit in the hot fire ash or heat it in the hot fire and have it as food item or chutney for 2-3 days or anytime of the day. It helps in suspension for adults. The best season to have is during October to April.

How to identify: Any eatable banana fruit is eatable but some are considered not delicious like the wild banana plant.

40. Botanical Name: *Nicotiana tabacum.*Linn. **[Family:** Solanaceae**], Common Name:** Tobacco.

Local Names - Maring: *Heelaknaa/ Sagonta.* **Manipuri:** *Utonglei.*

Diseases/Treatment: a) Fever, b) Joint-sprain.

Parts used: Leaf.

A) Heat the leaf on the fire and smash little bit and apply on the forehead or body areas. B) Crush the leaves and apply the paste on the sprain area by re-dressing it at each interval till cured.

How to identify: The plant is semi-shrub and grows up to 7-8 feet in high around tropical temperate zones. The leaves are big and rough. The flowers are like gramophone, big and are white to pinkish in colour. The locals dried the leaves and used as biddy or cigarettes.

41. Botanical Name: *Oroxylum indicum* (L.) Vent. **[Family:** Bignoniaceae**], Common Name:** Indian trumpet flower.

Local Names - Maring: *Thampak/ Khe / Shamba.* **Manipuri:** *Shamba.*

Diseases/Treatment: a) Piles, b) Tonsilitis/ Sore Throat/ Sinus.

Parts used: Bark, leaf.

a) Roast a bit of the fresh fruit on the fire and have it along with chutney regularly or alternately to cure the piles.

b) Boil some of the *Oroxylum indicum* bark or leaves with little salt (spring salt) and add little *Tekta* (*Lomba*) leaves and served as decoction half a glass twice a day after food for a week.

How to identify: The *Oroxylum indicum* (L.) Vent is a tree found grown in tropical temperate zone with big elongated leaf. The plant bears a long tree beans. The young fruit can be eaten by roasting, frying and is considered useful for other ailments.

42. Botanical Name: *Phlogacanthus thyrsiflorus* Nees. [**Family**: Acanthaceae], **Common Name**: Basak. **Local Names - Maring:** *Simlim/ Ramsingrim.* **Manipuri:** *Nongmangkha Angangba.*

Diseases/Treatment: Cough, Dry cough, Congestion of throat.

Parts used: Leaf.

Crush the leaves and mix with little honey and apply around the chest and inner earlobes and on the head.

One may also boil some leaves and have as decoction a glassful twice a day after food till relief.

(Collect some leaves of *Phlogacanthus thyrsiflorus, Adhatoda vasica, Zanthoxylum acanthopodium, ginger* and *Sitamasi* (white sugar cube) and boil it till blackish in colour and have it as syrup half a glass before food twice a day till relief. Do not eat red meat and fermented dry fish (*Ngari*) during medication. It is considered a very strong medicine for adults. For children, below five years two tea spoonfuls twice a day).

How to identify: This shrub plant is commonly grown in Manipur in the courtyards as fencing since it shows the perennial nature. There are different varieties but this yellow flower plant is commonly consumed as culinary.

43. Botanical Name: *Phyllanthus emblica* (L.), [**Family**: Labiateae], **Common Name**: Gooseberry.

Local Names - Maring: *Puklu-hei.* **Manipuri:** *Heigru.*

Diseases/Treatment: (a) Dry Cough/ Asthma (b) Headache, Hypertension (c) Sore Eyes.

Parts used: Fruit, bark.

For children, crush some amount of goose berries and mix the extracted liquid with little honey and give it before and after food three times a day till relief. For adults, to get immediate relief from headache and giddiness eat plenty of gooseberries and drink plenty of water. For those suffering from sore eyes, boil the gooseberry bark or root and splash the liquid on the eyes repeatedly while washing the face in the early morning, afternoon and bed time till cured. Or apply few drops of extracted gooseberry juice directly on the sore eyes 2-3 times a day till cured. It is believed to even cure cataract cases.

44. Botanical Name: *Plumeria acuminate* Ait. [**Family**: Apocynaceae], **Common Name**: Temple/Pagoda tree.

Local Names - Maring: *Leihou khongsang.* **Manipuri:** *Khage-leihou angouba.*

Diseases/Treatment: Malaria/ Fever.

Parts used: Bark.

Crush the dry bark into powder form and mix half a tea spoon in half a glass of water and take twice a day before each meal. The result is that the stool/ faeces will become black.

How to identify: This *Plumeria acuminate* or Temple/Pagoda tree is commonly grown in Manipur. The plant locally derives its name from *Khage/Kege leihou* literally meaning the 'Chinese-Thai aromatic flower', believed to have introduced the plant initially to the southern Manipur at Moirang in early days. It is grown in temperate tropical climate zone, at the family's courtyard.

45. Botanical Name: *Plumeria rubra* Linn. [**Family**: Apocynaceae], **Common Name**: Red Temple/Pagoda tree. **Local Names - Maring:** *Leihou-khongsang Anganba.* **Manipuri:** *Khage-leihou Angangba.*

Diseases/Treatment: Tonsil/ Malaria.

Parts used: Fruit/ Seed.

Dry the fruit/seed in the sun and when it turned black, cut into small pieces and eat a piece any time of the day. Store the small pieces in a tight container. The plant seldom bears such fruit/ seeds. (Remember, do not cut the fruit on any Saturdays).

How to identify: This *Plumeria rubra* or red temple tree is rarely found in Manipur. Of different red coloured varieties the fruit of the blood red colour flower plant is considered most valuable. Locally, the plant derived its name from *Khage/Kege leihou* literally to mean the 'Chinese-Thai aromatic flower' since it is believed that the flower was brought by them in early days during territorial expansion initially to southern Manipur at Moirang. It is grown in temperate climate zone especially around family's courtyard.

46. Botanical Name: *Psidium guajava* Linn. [**Family**: Myrtaceae], **Common Name**: Guava.**Local Names - Maring:** *Pungtol.* **Manipuri:** *Pongtol.*

Diseases/Treatment: Dysentery/ Diarrhoea.

Parts used: Leaf (Tender) / fruit.

Eat some tender leaves of any guava plant (or the unripe guava fruits) to help suspend the dysentery or diarrhoea. The leaves are little pungent and bitter.

47. Botanical Name: *Punica granalum* Linn. **[Family**: Onagraceae], **Common Name**: Pomegranate. **Local Names - Maring:** *Kapo-hei.* **Manipuri:** *Kaphoi.*

Diseases/Treatment: Dysentery/ Diarrhoea.

Parts used: Leaf, fruit.

Boil some leaves of *Punica granalum* and have as decoction a glass full 2-3 times a day. It helps in suspension.

48. Botanical Name: *Rauvolfia serpentine* Linn. **[Family:** Apocynaceae**], Common Name:** Indian snakeroot. **Local Names - Maring:** *Lappar marao/ Yenbum manbi* (Red flower). **Manipuri:** *Yenbum manbi.*

Diseases/Treatment: Epilepsy/ Asthma.

Parts used: Whole plant/ Root.

The roots or whole plant are boiled and served as decoction twice a day after food to cure epilepsy and asthma.

How to identify: The plant is not exactly the species of *Rauvolfia serpentine* Linn although, it belongs to the Apocynaceae family. This herbal evergreen plant grows in the deep tropical temperate forest zones especially, under the shade of big trees where there is lots of humid and moisture. The leaf looks like *Yenbum* plant and therefore is locally called as *Yenbum manbi*. The leaf is thick, smooth, shinny and slippery like plastic on both sides. The stalk and roots are hard and strong. The flower is red in colour and blooms in clusters at the tip of the branch. The high of the plant is normally 1 - 3 feet.

49. Botanical Name: *Rhus semialata* Murr. [**Family**: Anacardiaceae], **Common Name**: Wild varnish tree. **Local Names - Maring:** *Khongma hei.* **Manipuri:** *Heimang.*

Diseases/Treatment: Diarrhoea, Dysentery, Digestion.

Parts used: Fruits, Leaf.

Soaked some of the fruits in a glass of water by adding little common salt and served it to helps indigestion, diarrhoea, and dysentery. Chew some of the tender leave to fight against diarrhoea.

How to identify: The plant bears fruit in winter session in clusters of bunches. The tender fruits are pinkish and while maturing it is filled with acidic snow white turns brown when ripen. The fruit has the taste of nitric acid. The leaf is little rough on both sides and has clumsy taste when chewed.

50. Botanical Name: *Saccharum officinarum* L. [**Family**: Poaceae], **Common Name**: Red sugarcane. **Local Names - Maring:** *Masu marao/ Chu-ngang.* **Manipuri:** *Chu-ngang.*

Diseases/Treatment: Jaundice (*Thongngak*).

Parts used: Stem.

Take a glassful of fresh (red) sugarcane juice daily anytime of the day till it is cured. For children: Apply the cooked mustard oil with garlic on the child's body and exposed in the sunlight. Give half a glass of fresh sugarcane juice three times daily till it helps cured. (Do not give *Ngari* the local fermented dry fish, meat, and chilly, but simple boiled papaya curry during medication).

51. Botanical Name: *Schefflera arboricola* (Hayata) Kanehira. [**Family**: Araliaceae], **Common Name**: Dwarf umbrella plant. **Local Names - Maring:** *Lam-om.* **Manipuri:** *Utangbi.*

Diseases/Treatment: a) Sprain/ Fracture of hands and legs, b) Infertile women.

Parts used: Leaf.

c) Boil with little common salt some leaves of it and make paste by adding little honey and apply on the sprain or fracture part of the leg or hand.

d) It is even use in ritual for infertile women.

How to identify: This is a big climber plant that grew by supporting on bigger trees. It belongs to the family of Gesneriaceae plant species. The fruit when ripe are yellow to orange colour bears like wild berries and likes by birds. It is used even in ritual for infertile women. It is commonly grown in tropical temperate forest areas. The stalk has six leaves each at the end of the tip like flower petals. The leaves are thick, soft and shiny. It belongs to hardwood family.

52. Botanical Name: *Smilax ovalifolia* Roxb. [**Family**: Liliaceae], **Common Name**: Kumarika/ Aushbah). **Local Names - Maring:** *Truntu / Kangra.* **Manipuri:** *Kharam mana/ Kwa-manbi.*

Diseases/Treatment: Kidney/ Stone problems.

Parts used: Leaf.

Boil the leaves in certain amount of water with Sitamasi (white sugar cube) and serve as decoction a glassful three times a day till relief or cured. The body will swell up a bit and drains out any negative element of body in the form of extracted liquid or urine. (One should not take pumpkin, yam and *ngari* (local fermented dry fish) during the medication).

How to identify: This is a climber plant grown in the tropical forest areas. There are different varieties. The plant has small thorns on the stem and at knots. The leaves are hard and shiny smooth. The barks are hard.

53. Botanical Name: *Solamum virginuanum* Linn. [**Family**: Solanaceae], **Common Name**: Yellow berried nightshade. **Local Names - Maring:** *Samtrok-kha.* **Manipuri:** *Leibungkhang.*

Diseases/Treatment: Headache/ Toothache.

Parts used: Fruit.

Crush some amount of the *Solamum virginuanum* fruits and add two spoons of honey and have it till the pain is gone.

54. Botanical Name: *Syzigium fruticosum* DC. [**Family**: Myrtaceae], **Common Name**: …….. **Local Names - Maring:** *Heinou-manbi.* **Manipuri:** *Heinouman / Tomba-heinou.*

Diseases/Treatment: Fever (especially for children).

Parts used: Leaf.

Some amount of *Syzigium fruticosum* leaves are boiled and is taken (sponge) bath with it when fever is little down. Or use the boiled water for sponge bath. (This is for external use only. Do not drink the boiled leaves water).

How to identify: The shrub plant bears like a retarded tiny mangoes and has the smell of mango with small white flowers that is why the locals called it as *Tomba Heinou* (lit. Junior mango). Some leaves are yellowish to deep pink to red colour, often grown in sloppy dry tropical forest areas. It is different from *Syzygium* cumini (Jamun) plant.

55. Botanical Name: *Tamarindus indicus* (L.) [**Family**: Caesalpiniaceae], **Common Name**: Tamarind. **Local Names - Maring:** *Mangke.* **Manipuri:** *Mangge.*

Diseases/Treatment: Bee Sting.

Parts used: Seed.

Cut the tamarind seed into half and apply the white part on the bee stung area and lightly bandage it. This helps reduce the pain.

56. Botanical Name: *Urena lobata* Linn. **[Family:** Malvaceae**], Common Name:** Ceasarweed. **Local Names - Maring:** *Chumphal.* **Manipuri:** *Asikhangla.*

Diseases/Treatment: Stone cases Kidney and gall bladder. **Parts used**: Leaf / stem.Boil the leaves and stems and have as decoction a glassful three times a day before food till cured.

How to identify: This herbal plant grow wild in the wasteland. The leaves are thick, coarse and rough. The stem are strong to break. The flower is small and pinkish in colour.

57. Botanical Name: *Vitex negundo* Linn. [**Family**: Verbenaceae], **Common Name**: Chaste tree. **Local Names - Maring:** *Warek-lou.* **Manipuri:** *Urik-shibi.*

Diseases/Treatment: Piles.

Parts used: Leaf (Whole plant).

Burn the dry leaves and collect the ashes and apply on one's anus 3-4 times a day. Or boil some fresh leaves and serve as decoction a glass full before each meal. (Do not eat red meat, chicken and egg during medication). Result is known within 3/4 days.

How to identify: This shrub plant bears bluish white flowers from small buds in clusters. The leaves are small and smooth and have little grass smell. It grows around dirt or waste areas.

58. Botanical Name: *Xylosma longifolia* Clos. [**Family**: Salicaceae / Flacourtiaceae], **Common Name**: Dandal (Hindi). **Local Names - Maring:** *Nungshang-panbi.* **Manipuri:** *Nungleishang.*

Diseases/Treatment: a) Piles, b) Menstrual cycle problem, c) Paralysis in men.

Parts used: Leaf.

a) Boil some quantity of the *Xylosma longifolia* leaves and have 3-4 tea spoonfuls twice a day after food till it is cured.

How to identify: This shrub plant is spiky. The flower is yellow with lots of pistils bloom in bunches at every knot leaf. The leaves are shiny and smooth though tender leave are reddish in colour. The variety is somewhat different from Dandal (Hindi) plant.

59. Botanical Name: *Zanthoxylum acanthopodium* [**Family**: Rutaceae], **Common Name**: Pricky winged leaf/ Darmar/ Sheguan pepper.

Local Names - Maring: *Singdi.* **Manipuri:** *Mukthrubi.*

Diseases/Treatment: Gas Formation.

Parts used: Leaf and Seed.

The dry brown seeds' covering are crush into powdered form and about 5 mg is mixed in half a glass of warm water and taken for 2-3 days, or may take along with other food items. It acts as anti-oxidant agent. (Use only the brown seed covering by removing the black seeds).

How to identify: The fresh leaves and dry brown seed covering of this prickly shrub plant are widely used in meat curry by most North-east tribes of India rather than as medicines. The leave has slight pungent aromatic smell while the seeds have a strong pepper taste. The seed while ripening is red and when dry is dark brown. It is grown in tropical climate areas. The taste is better than black pepper.

60. Botanical Name: *Zingiber spectabile* **[Family:** Zingerberaceae**], Common Name:** Thai Ruby Ginger (Wild ginger) **Local Names - Maring:** *Lamsing* **Manipuri:** *Lamsing.*

Diseases/Treatment: Gas formation, flatulent, stomach upset.

Parts used: Rhizome.

Crush the rhizome and add little honey and have as syrup two tea spoonfuls thrice a day till cure.

How to identify: This plant belongs to the family of Zingerberaceae and grows like wild ginger with few stalks. Therefore, the locals called it as *Lamsing* literally 'wild ginger'. The rhizomes are small and have strong aromatic smell. There are two varieties- male and female; male grows up straight and the leaves are small and the female plant is short and has broad leaves. The species also appears to be closer to *Zingiber montanum* with an average high of 1-1.5 metre.

CHAPTER - 3

◆◆◆◆◆

The Practice of Ethnomedicines among the Maring Tribe

1. Botanical Name: Local Names - Maring: *Akdirung.* **Manipuri:**

Diseases/Treatment: Gas formation in stomach. **Parts used**: Rhizome.

Crush the dry rhizome of this creeper plant in powdered form and mix a little of cow's bile juice and have as syrup two tea spoonful twice a day before each food for three days. The effectiveness is indicated by constant excretion of urine and faeces. (They believed that even patient troubled by contact with evil spirits also helps cured).

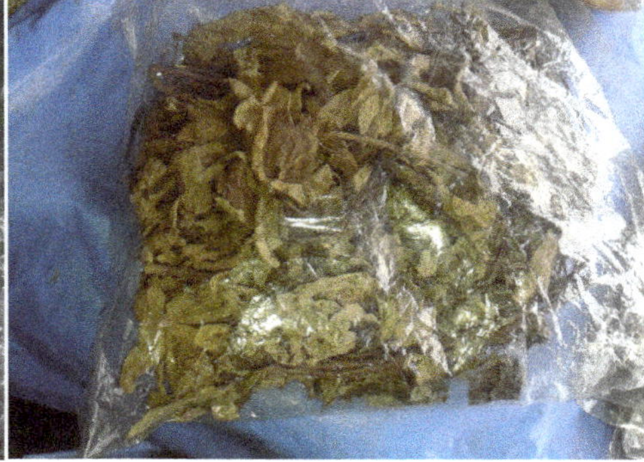

2. Botanical Name: Local Names - Maring: *Benacha.* **Manipuri:**

Diseases/Treatment: Blood Purifier/ Menstrual/ Urinal Problems.

Parts used: Whole plant.

Boil the whole creeper plant (fresh or dry) till the water is deep red and serve as decoction half a glass twice a day regularly before each food for a week. It helps in rejuvenating the internal body organs especially for post-natal women.

How to identify: This is a creeper plant grown in the deep interior tropical temperate forest areas. The dry plants wrapped in packages are sold by local vendors at Kakching Lamkai, Manipur.

3. Botanical Name:

Local Names - Maring: *Burmanu.* **Manipuri:**

Diseases/Treatment: a) Headache/ Migrants, Giddiness, b) Hypertension.

Parts used: Leaf.

a) Crush this tree leaves and apply on the top of the head or forehead. One may also mixed with little rice starch for soothing effect. Repeat this process 2-3 times a day till relief.

b) Boil some leaves and take bath with the extract water for a week regularly. This is especially for women suffering from hypertension and strong migrants/ headache.

How to identify: This shrub plant is widely grown in the local tropical forest. It belongs to hardwood family.

4. Botanical Name:

Local Names - Maring: *Chinghaya mathon.* **Manipuri:**

Diseases/Treatment: Chest pain.

Parts used: Leaf.

Crush the shrub leaves and apply the paste on the chest and other body parts till relief by repeating it at each interval.

How to Identify: The plant's fruit is like pea wildly grown in the roadsides and wasteland areas. It is not eatable.

5. Botanical Name:

Local Names - Maring: *Hatia.* **Manipuri:**

Diseases/Treatment: Kidney stone cases.

Parts used: Bark.

Boil the dry barks in certain amount of water and have as decoction a glassful thrice a day regularly for two weeks.

How to identify: It is found sold by Khoibu Maring tribe in the Kakching Lamkhai market, near Pallel.

6. Botanical Name:

Local Names - Maring: *Kaksilou.* **Manipuri:**

Diseases/Treatment: Urinal problem/ Stone cases in kidney and gall bladder.

Parts used: Leaf.

Boil some leaves till deep red and serve as decoction a glassful thrice a day till relief.

How to identify: The flower this shrub plant is bluish in colour and blooms in bundles. The tender leaves are eaten as chutney by the locals.

7. Botanical Name:

Local Names - Maring: *Kaporlou.* **Manipuri:** *Kang panbi.*

Diseases/Treatment: a) Body swelling (Internal boils), b) Allergy.

Parts used: Leaf.

a) Crush the fresh leaves and apply on the body swelling area for a night. Repeat this process for three or four days till cured. b) Boil the leaves and serve as syrup three tea spoonfuls thrice a day for a week.

How to identify: This *Kaporlou* is a climber plant. The leaves are elongated, smooth and shiny, where the stems and branches are hard.

8. Botanical Name:

Local Names - Maring: *Khamting.* **Manipuri:**

Diseases/Treatment: Measles, chicken pox.

Parts used: Root/ Rhizome.

Crush with little water and add little bear's bile juice of this herb plant and apply on the whole body. This is especially for children suffering from measles and chicken pox.

9. Botanical Name:

Local Names - Maring: *Khanglung.* **Manipuri:**

Diseases/Treatment: Sprain/ Fracture of hands and legs.

Parts used: Leaf.

Crush the leaves this climber plant and apply the paste on the sprain or fracture parts by wrapping it for a day or two.

10. Botanical Name: Marigold

Local Names - Maring: *Kilnambo / Sanarei.* **Manipuri:** Sanarei

Diseases/Treatment: High fever. **Parts used**: Leaf.

Boil the leaves and have three times a day as decoction half a glass till cured.

11. Botanical Name:

Local Names - Maring: *Kinneem.* **Manipuri:** *Kongpe mana / Kinneem.*

Diseases/Treatment: Fever.

Parts used: Leaf.

Boil some leaves of *Kinneem* in a liter of water and drink a glassfull twice a day after food till the fever is reduced. (The leafs looks like Jumun leafs).

How to identify: It is a shrub plant that grows in the tropical forest. The leaves are thick and not so smooth.

12. Botanical Name:

Local Names - Maring: *Kongkohui.* **Manipuri:**

Diseases/Treatment: a) Diabetes b) Stone cases in kidney. **Parts used**: Roots, Leaf.

a) Boil some roots of this shrub plant with some water and have as syrup two tea spoonfuls three times a day till cured.

b) Boil the leaves with some Sitamasi sugar (white sugar cube) and have a syrup three teaspoonful three times a day. The urine excreted will be black in colour.

13. Botanical Name:

Local Names - Maring: *Kuldang.* **Manipuri:**

Diseases/Treatment: Snake bite.

Parts used: Leaf.

Crush the male plant leaves of this shrub plant and apply the paste in bitten area.

14. Botanical Name:

Local Names - Maring: *Kulbirui.* **Manipuri:** *Wangbarel.*

Diseases/Treatment: Heart and Kidney Failure Problems.

Parts used: Bark, stem.

Boiled half a Kilogram of the climber bark in about 5 litres of water till it turns deep red. Bath with the extracted boiled water or apply regularly on the chest

and abdomen pain areas any time of the day regularly for a week or more. This medication is exclusively for external purposes only. Therefore, do not drink the extracted liquid.

How to identify: This *Wanbarel* is a climber plant grown around a sacred grove in the deep interior forest. The climber plant is locally known as *Wangbarel* wife's medicine since it cured one of the legendary king's wives of Moirang, Manipur. It is claimed to grow in tropical temperate areas. It is found in Chakpikarong, Chandel district. (Source: Menai Maring of Pallel).

15. Botanical Name:

Local Names - Maring: *Kursi.*

Manipuri:

Diseases/Treatment: Cuts, wounds, sprain.

Parts used: *Fruit.*

Crush the fruits of this creeper plant into powder form and apply on the cuts and wounds areas till cured by changing the bandage.

16. Botanical Name:

Local Names - Maring: *Lamkumlou.*

Manipuri: *Chakmang mana.*

Diseases/Treatment: Female blood purifier (Post-natal woman).

Parts used: Roots, Leaf.

Boil and have two tea spoonfuls as syrup three times a day of this herb plant.

17. Botanical Name:

Local Names - Maring: *Loumeh.*

Manipuri:

Diseases/Treatment: Diarrhoea, Dysentery, Typhoid.

Parts used: Root.

Boil the roots of this herb plant till deep red and serve as decoction a glassfull three times a day till cured. (Do not take chilly, fruits, prawn, eel during this medication). The dry roots can also be used.

18. Botanical Name:

Local Names - Maring: *Loulurpui.* **Manipuri:** *Hameng sampakpi*

Diseases/Treatment: Typhoid.

Parts used: Leaf.

Boil the leaves of this herb plant and have as decoction half a glass twice a day till cured.

How to identify: This herbal plant wildly grows in the dry waste land areas. The leaves are rough, hairy, spiky, small and sticky when crushed. Normally it grows upto 3-4 feet high.

19. Botanical Name:

Local Names - Maring: *Louthaither.* **Manipuri:**

Diseases/Treatment: High fever/ tooth ache.

Parts used: Roots.

Boil the roots of this herb plant and have as syrup two spoonfuls twice a day till cured.

20. Botanical Name:

Local Names - Maring: *Lungpaar.*

Manipuri: *Nungthambal.*

Diseases/Treatment: Cancer. **Parts used**: Stem/ Flower.

Smear the stem/ flower on a clean hard stone by mixing with little water and collect the extracted liquid with cotton and apply it on the cancer part and bandage it well. Repeat this formula after two days and continue for 3-4 months. It should be done before chemotherapy treatment. This is for external used only. Other items and ingredients are also used.

How to identify: This herbal plant *Nungthambal* (lit. stone lotus) species is considered very rare and valuable. It is usually found grown only above the rock like mushroom plant or like small red lotus in the deep forest with leaves or without leaves especially, in the months of May-June. The flower is white in colour, thick and very hard. It is claimed to have found in Laibi Maring village in Tengnoupal district.

21. Botanical Name:

Local Names - Maring: *Ngalpho.* **Manipuri:**

Diseases/Treatment: Fever.

Parts used: Leaf.

Crush the leaves and apply the paste on the forehead and other body parts to reduce the fever.

22. Botanical Name:

Local Names - Maring: *Nunghoak Sirik.*

Manipuri:

Diseases/Treatment: Leprosy.

Parts used: Root.

Extracted oil of the roots is applied on the whole body parts time to time till cured.

23. Botanical Name:

Local Names - Maring: *Pankhok.*

Manipuri:

Diseases/Treatment: Body swelling.

Parts used: Rhizome.

Crushed the rhizome and add the extracted liquid of the thunder-bolt stone and apply on the swelling body parts. This is a kind of wild water yam.

24. Botanical Name:

Local Names - Maring: *Rultumlou.*

Manipuri:

Diseases/Treatment: Irregular menstrual problems, Blood purifier (Post-natal woman). **Parts used**: Whole plant.

Boil the whole creeper plant and serve as decoction a glassful by mixing a spoonful of honey in a glass and have three times a day for a week. The boiled extracted liquid is red in colour. This herbal medicine is especially recommended for post-natal women who suffered from irregular menstrual or for irregularity.

25. Botanical Name:

Local Names - Maring: *Sakon hidak trung/ Mickyekpaar.* **Manipuri:** *Kothap mana.*

Diseases/Treatment: Removal of bad blood, poisonous insect bites, boil.

Parts used: Leaf.

Crush the leaves and apply the paste on the swelling area of the bad blood, or bite of any poisonous insect bites or boil area till cure.

How to identify: This shrub plant grows around the swampy areas. It has large coarse leaves. The flowers are large like gramophone and are white flowers in colour. The plant grows upto a high of about 6-8 feet tall.

26. Botanical Name:

Local Names - Maring: *Salda wa.* **Manipuri:** *Nat wa/va.*

Diseases/Treatment: Tonsillitis.

Parts used: Leaf.

Boil some of this bamboo leaves and have as decoction three times a day till cured the tonsillitis. During recovery one would feel thirsty for water.

How to identify: The locals can easily identify this variety of bamboo species, besides others important plants.

27. Botanical Name:

Local Names - Maring: *Taangkha/ Taangkinu.* **Manipuri:** *Yanungkha.*

Diseases/Treatment: Tonsil/ Sore Throat Problem.

Parts used: Leaf, whole plant.

Boil the leaves or some amount of this herb *Yanungkha* whole plant in about three litres of water and have as decoction a glassful twice a day after each meal. One may also have it in raw fresh form till it helps cure. (Do not give to girls who are weak and suffering from leukaemia). Some claimed it helps fight against cancer, boil types, stomach ailments too.

How to identify: This herb plant grows in the cold temperate areas. The normal high of the plant is about 2-3 feet tall.

28. Botanical Name:

Local Names - Maring: *Thingkangphu.* **Manipuri:**

Diseases/Treatment: a) Liver problem (alcoholic), Ulcer, b) Gastritis and Stomach problems. **Parts used**: Bark, Leaf.

a) Boil the bark of this tree plant with some amount of water and serve as decoction twice a day for a week to help relief from chronic ulcer, liver and stomach problems. This is especially recommended for patient suffering from alcoholism.

b) Boil the leaves in about three litres of water and served as decoction half a glassful twice a day before food till it cures. (Do not eat spicy things like chilly, ginger, etc. during medication).

How to identify: The *Thingkangkhu* is a tree. The leaves are elongatedly large and slightly coarse.

29. Botanical Name:

Local Names - Maring: *Thingphungcho.* **Manipuri:**

Diseases/Treatment: Kidney and Gall badder Stone Cases.

Parts used: Leaf/ Bark.

Boil the leaves in about five litres of water and served as decoction a glassful thrice a day before food till it is cures. This species is believed to serve as anti-oxidant too. ***How to identify:*** The dry leaves wrapped in packages are sold by local vendors at Kakching Lamkai, Manipur.

30. Botanical Name:

Local Names - Maring: *Thruna.* **Manipuri:**

Diseases/Treatment: Urinal problem/ Stone Kidney cases.

Parts used: Leaf.

Boil some amount of the leaves till deep red and drink as decoction three times a day till relief of pain. The urine will flow freely.

How to identify: The small yellow flowers are eaten by the locals as chutney.

31. Botanical Name:

Local Names - Maring: *Tumpina.* **Manipuri:**

Diseases/Treatment: Boil. **Parts used**: Leaf.

Crush the leaves and apply the paste on the boil area by repeating at certain intervals.

32. Botanical Name:

Local Names - Maring: *Waa-aryam.* **Manipuri:**

Diseases/Treatment: Body-ache, Weakness, Post-natal.

Parts used: Leaf.

Take bath with the boiled fresh leaves and one may used as bam with the hot extract water regular till relief.

How to identify: This is a climber plant. The flower is reddish and pinkish in colour. The leaves are small rounded and smooth. Some people grew near the houses as a decorative plant.

33. Botanical Name:

Local Names - Maring: *Yaili.* **Manipuri:** *Ngamu yai.*

Diseases/Treatment: Scabies.

Parts used: Roots.

Crush the roots of this climber plant and apply the paste on the scabies area three-four times a day till cured. The crush roots and barks are also used for fishing in the river cleavages by locals.

How to identify: This is a climber plant found grown in deep temperate forest. The tender leaves are yellowish to brown and become dark green when mature.

34. Botanical Name:

Local Names - Maring: *Yasikhong*. **Manipuri:**

Diseases/Treatment: Snake bite, dog bite.

Parts used: Rhizome/ Root.

Crush the rhizome of this Yasikhong climber and apply the paste on the bitten part till cured.

35. Botanical Name:

Local Names - Maring: *Youlou*. **Manipuri:**

Diseases/Treatment: Urinal Problem/ Stone kidney cases.

Parts used: Roots.

Boil some leaves with Sitamasi (Crystal cane sugar) and serve as decoction a glassful three times a day till relief. It is colourless and has no smell. (It is small herbal plant with small leaves and small white flowers).

Other Forms of Indigenous Medicines Used by Maring Tribe

1. Scientific Name: Crystal cane sugar.

Local Names - Maring: *Sitamasi / Chini matum.* **Manipuri:** *Sitamasi (Misri* in Assamese*).*

This *sitamasi* or sugar crystal is available in local market and is an important ingredient. Some amount of it is added while boiling certain ethno-medicinal plants like leaves, barks or creeper or climber and often served as decoction.

2. Scientific Name: *Rice*

Local Names - Maring: *Sai.* **Manipuri:** *Cheng.*

Diseases/Treatment: Dog bite.

Parts used: Rice paste.

Chew the raw rice and apply the paste on the dog's bitten part and place a silver coin on it.

3. Scientific Name: *Fermented local rice-beer.*

Maring Name: *Khaji /Chakthamwa.* **Manipuri Name:** *Athing mapang.*

Diseases/Treatment: Urinal/ Kidney Stone Problem.

Parts used: Residual rice beer.

Drink three-four mugs of sticky fermented rice beer in a day for atleast a month or more. It helps in the excretion along with the urine.

4. **Scientific Name:** *Kitri (Warhogs).*

Local Names - Maring: *Lam-ok manbi.***Manipuri:** *Lam-ok (Kitri).*

Diseases/Treatment: Leprosy (*Richi* in Maring).

Parts used: Blood, Whole body (meat).

The person infected with leprosy may apply the blood of this unique animal regularly and also eat the cooked meat. It is believe that it cures within a weeks' time.

How to identify: A unique animal with red and bluish slicken plume that looks like a wild boar and has a limb/ hoof like a dog. They live six month in a burrow and six months outside on the surface in a year.

5. **Scientific Name:** *Lethocerusamericanus* [Belostomatidae], Giant Water Bug (Lake's species).

Maring Name: *Nausek.* **Manipuri Name:** *Nausek achouba.*

Diseases/Treatment: Snake Bite/ Bitten by any Poisonous Reptiles.

Parts used: Head.

Get the head of the lake's insect *Lethocerusamericanus* or Giant Water Bug species and stuck its head at the snake bitten part. It stuck and sucks out all the venoms automatically till it is clear. The *Nausek* head will drop once it finishes sucking it.

How to identify: This giant water bug's head is considered very valuable for many local medicine men for the treatment of snake bites and other poisonous bites. According to some local healers, they said that when a snake threatens the territory of giant water bugs they fly and sting upon the snake's head and killed it. Even the dry head of the giant water bug is preserved by some as a valuable item.

6. **Scientific Name:** *Planorbella trivolvis* [Planorbella], Gastropod mollusc (Freshwater snail). **Maring Name:** *Masol.* **Manipuri Name:** *Tharoi.*

Diseases/Treatment: Snake Bite.

Parts used: Mucus membrane (body).

Get a fresh snail and put the head around the bitten part without disturbing it for a while. It is believe to suck the venoms out.

How to identify: The recommended snail of fresh water snails that are eatable it may be either short or long knot type. But one has to be very careful which are not eatable.

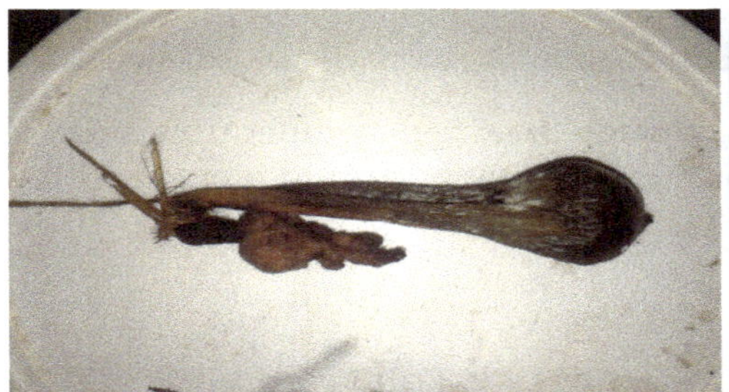

7. **Scientific Name:** *Pig's Bile.*

Local Names Maring: *Mit.* **Manipuri:** *Masingkha.*

Diseases/Treatment: Measles, Chicken pox.

Parts Used: Bile juice

Python's bile, Beer's bile, Black monkey's bile, and crow's bile all these bile mixed together with little water is applied on the body to help cure the measles and even chicken pox. (Most of the animals' bile juice dry or fresh is considered very usefull for local medicine man in the preparation of their indigenous medicines for the treatment of various diseases and infection).

How to identify: The bile sack is located attached to the liver and spleen of an animal's organ, which is greenish in colour. Because of its bitterness such bile's is never added in curry.

8. **Scientific Name:** *Thunder-bolt stone.*

Local Names - Maring: *Thaanmuna rei/ Maran thlung* **Manipuri:** *Braja nung.*

Diseases/Treatment: Fever, Stomach upset.

Parts used: Extracted liquid.

The thunderbolt stone is rub against the hard stone by adding little water and the extracted liquid is collected with cotton and apply on the forehead, body and is

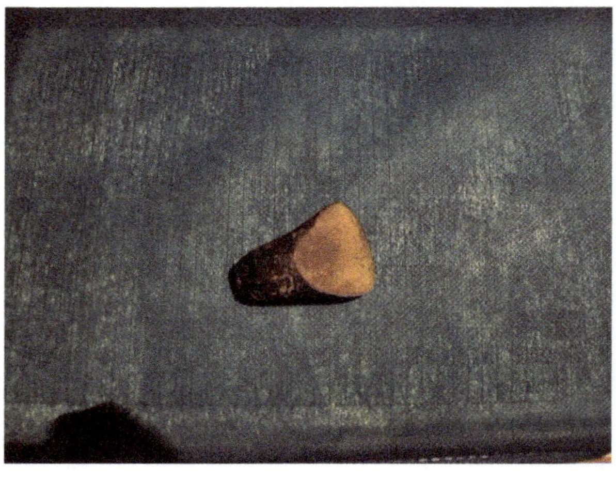

also suggested to drink little to cure the fever and stomach upset when believe to have caused by evil spirits.

How to identify: Such thunderbolt stones are meteorite stones dropped down during thunder and storms of rainy season. Some tribals term it as heaven's stone or God's stone. Such thunderbolt stones are considered rich in certain minerals since, it is part of a rock substance. Often such thunderbolt stones are in axe shaped, although the size may differ from small to large. It comes in different colours, some are slightly greyish to brown and it is very hard.

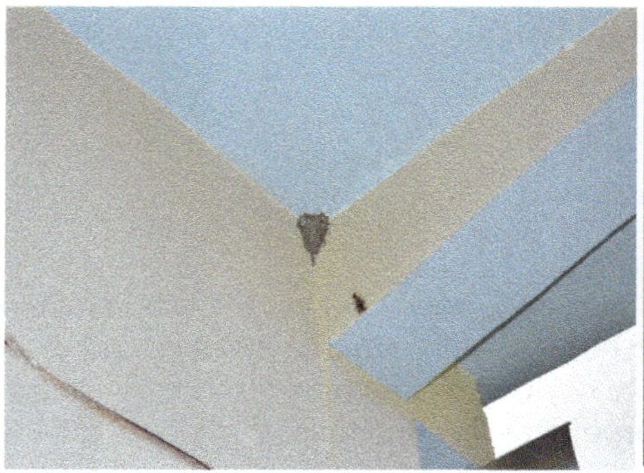

9. Scientific Name: *Vespula germanica* F. [Arthropoda], House wasp.

Maring Name: *Chim/thlei khui* (*Leibak khoi*). **Manipuri Name:** *Leibak khoi.*

Diseases/Treatment: Mumps (Type).

Parts used: Nest mud.

Mixed the mud collected from the house wasp with saliva or with little water and apply it on the mumps area 3-4 times a day till heal. (It cures but the reason is unknown said a respondent).

How to identify: Any house wasp mud is considered good. There is no specific that this type of house wasp mud is better or less in quality so far according to informants.

CHAPTER - 4

◆◆◆◆◆

Analysis of the Data

Table 1: Indigenous Medicinal Plants of Maring (Khoibu) Tribe

Sl. No.	Botanical Name [Family] Common Name	Maring Name [H, Sh, T, C, Cl]	Manipuri Names [(S)/ (P)]	Parts Used [O / E]	Ailments and Diseases: Treatment Methods and its Dosages -
1.	*Adhatoda vasica* Nees. [Acanthaceae] Malabar nut	*Triptung-ngou* [Sh]	*Nongmangkha angouba* [P]	Leaf [O/E]	(a) Fever (high), (b) Cough. a) Crush the leaf and collect the liquid and add little honey and serve a spoonful twice a day. b) Crush leaves are mixed with kerosene and apply at the back of the earlobe or boil the leaves and may also bath with it to relief from cough.
2.	*Aeschynanthus hookeri* C.B.Clarke, [Gesneriaceae]	*Anlikli* [Sh]	*Utangbi* [S]	Leaf, bark [O]	Diabetes – A handful of this *Utangbi* leaves are boil and served twice as decoction one glass in empty stomach before each meal till the diabetes is cured.
3.	*Agaricus campestris* [Algae] (Meadow/Ground mushroom)	*Thrai-meetlung/ Leibak-marum* (Algae)	*Leibak-marum* [S]	Stem [E]	Burn – Apply the extracted liquid of the ground mushroom or the powder obtained on the burned part.
4.	*Ageratum conyzoides* Linn. [Asteraceae] Goat weed	*Yanglou* [H]	*Khongjai napi/ Lounamba* [S]	Leaf [E]	Cuts, wounds - Crush some fresh leaves and apply it on the cuts or wound areas.

5.	*Albizia myriophylla* Benth. [Leguminosae] Little-leaf sensitive-briars	*Thre-lou/ Silip yangyu* [Cl]	*Yanglee* [S]	Leaf [O] Bark, Root [E]	a) Urinary/ Kidney/ gall bladder Stone Cases, b) Dog Bites - a) Boil some leaves of *Yanglee* with *sitamasi* (white sugar crystal) and have 3/4 spoonful and take 3-4 times a day. The urine will be black in colour later if taken. b) Chew the *Yanglee* bark or root with little raw rice and apply the paste immediately at the bitten part.
6.	*Alocasia macrorrhiza* (Linn.) Schott. [Araceae] Giant taro	*Madung/ Lam baal bo* [H]	*Hongngo* [P]	Stem [E]	Bee sting - Cut the stem and immediately apply the extracted liquid on the bee stung area. One should be very careful not to touch the extracted liquid with hand for it is very itchy.
7.	*Alpinia allughas* Roscoe [Zingiberaceae] Galangal	*Puleimal* [Sh]	*Puleimanbi* [S]	Rhizome/ Root [O]	Gas Formation (Flatulence) – Crush the roots (rhizome) and mix with gooseberry and little honey and take one tea spoonful as syrup after food whenever thirsty.
8.	*Alpinia galangal* Linn. Willd. [Zingiberaceae] Greater galangal	*Ramrhou* [H]	*Kanghu* [S]	Rhizome [E]	Piles – Small amount of *Kanghu* is crushed along with some tobacco leaves and is then inserted inside one's anus. Do it twice a day till it is cured.
9.	*Antidesma acidum* Retz. [Phyllanthaceae] Amita	*Antuitik/ Ram Ansur* [Sh]	*Ching-yensil.* [P]	Leaf [O]	Arthritis/ Rheumatism (Joints pain) - Boil the leaves of *Oxalis corniculata* and served as decoction a glass full three times a day till cure. It may also be served as culinary item. (Fruit is black in colour when ripen).

10.	*Areca catechu* L. [Palmaceae] (palms) Areca nut/ Betel nut	*Kongkwai/ Kom kwa* [T]	*Kom kwai* [S]	Seed [E]	Irregular Menstrual Problems – Spelt upon the beetle nut and cut into half and throw away. (This involves use of magical charm while breaking).
11.	*Azadiracta indica* A. Juss [Meliaceae] Neem plant	*Neemtrung* [T]	*Neem* [S]	Leaf [O]	Malaria - The crushed leaves is mix with little water and about 2-3 tea spoonful is taken twice a day before each meal. (Do not give to those who are weak/ unhealthy/ pregnant or those suffering from leukaemia since it is very bitter).
12.	*Benincasa hispida* Thunb. Cogn. [Cucurbitaceae] Ash-gourd/ Winter melon	*Anmahei-angou (Kulbi),* [C]	*Torbot* [S]	Fruit [E]	a) Bear Bites, b) Tiger Bites – a) Apply immediately the paste of *Benincasa hispida* in the infected area. Then sprinkled the powdered *Kursi* seeds upon the wounded area. b) Peel off the green covering and sliced the *Benincasa hispida* fruits and apply the paste at the bitten part by wrapping it with a cloth.
13.	*Blumea balsamifera* D.C. [Asteraceae] Elumea or Nagal Camphor	*Langthrei* [H]	*Langthrei* [S]	Leaf [O]	Stomach Burning Sensation – Crush the leaves into paste form and mix with little water in a glass and drink it immediately anytime till it cures. Or one may eat few leaves and drink some water immediately to get relief soon.
14.	*Carica papaya* L. [Caricaceae]- Papaya	*Tengshangbo/ Khnachang hei* [Sh]	*Awathabi* [P]	Fruit [E]	Ulcer (Chronic Ulcer) - Boil the unripe papaya and have regularly as culinary or anytime. One may also have the ripe papaya regularly anytime till it cures.

15.	*Centella asiatica* Linn. [Apiaceae] Indian penny wort	*Anlaiphun* [H]	*Peruk* [S]	Whole plant, leaf [O]	a) Sore Throat, b) Hypertension – a) Boil certain amount of the whole plant in about 2 litres of water and have as decoction a glassful 3 times a day. Repeat it till is cures. b) One may have the fresh leaves regularly as culinary item to reduce hypertension or high blood pressure.
16.	*Chenopodium ambrosioides* Linn. [Chenopodiaceae] Wormseed / Jesuit's tea / tree.	*Kerosene-numbi* [H]	*Kerosene/ Tersing numbi* [S]	Leaf [E]	Worms in babies – Crush the leaves and apply on the baby's body as balm this will repel the intestinal worms. Repeat for three-four days.
17.	*Cinnamomum camphora* (Linn.) Nees & Eberm. [Lauraceae] Camphor tree	*Kapot* [Sh]	*Karpur pambi* [S]	Leaf [O]	Post-natal women, Anaemia – The leaves are boiled and taken bath by post-natal delivery woman. It also helps relief body ache and from annemea.
18.	*Cinnamomum zeylanicum* Breyn. [Lauraceae] Ceylon cinnamon	*Ushingsha* [T]	*Ushingsha* [P]	Bark [O]	Cough, Tonsillitis - Boil little amount of the dry *Cinnamomum zeylanicum* bark in about half a litre of water for sometime and add little Sitamasi (Crystal cane sugar) and served as decoction three times a day after food till relief.
19.	*Colocasia esculenta* (L.) Schott [Araceae] Yam	*Baal* [H]	*Pan* [S]	Tuber/ bulb [E]	Spike on Heel – Cut the yam and paste on it or bandage around the heel for three- four days.

20.	*Curcuma caesia* Roxb. [Zingiberaceae]- Black turmeric/ Black zedoary (Ginseng type)	*Aidai* [H]	*Yaimu* [S]	Rhizome [O]	a) Vomiting of Blood, b) Menstrual Cycle problem, c) Child Indigestion. a) The crushed black turmeric is mixed with few drops of water and have a half glassful before each food twice a day. The stool will be diffused immediately with blood. Repeat it till it helps cured. b) Crush certain amount of *Yaimu* and mix with little water and take 3-4 tea spoonfuls twice a day for a week before food. c) Crush some amount of Yaimu, collect the liquid with cotton and mix with little water. Give half tea spoon for child below 3 years and 1 tea spoonful above 3 years old. One may also apply the crush paste on the child's body. The child faeces will be very smelly.
21.	*Curcuma C. zedoaria*. Roscoe. [Zingiberaceae] Wild white turmeric/ Zedoaria/ Amba Haldi	*Aitonsan Angouba* [H]	*Yai-ngang Angouba* [S]	Rhizome [E]	Pigmentation/ Black/ Dark Spot/ Pimples – Crush the wild red turmeric and mix with little mustard oil and apply the paste regularly on the area before bedtime.
22.	*Curcuma longa* Linn. [Zingiberaceae] Turmeric	*Aitonsan* [H]	*Yai-ngang* [S]	Rhizome [E]	Cuts/ Wounds – Crush the turmeric and apply immediately on the cuts. It helps in blood coagulation and heals fast.
23.	Cymbopogon citratus (DC.) Stapf [Poaceae] Lemon grass	Lemon grass [H]	Lemon grass [S]	Leaf [O]	Sinusitis – Smash the leaves and apply the juice. It is also use for ringworm problem. Boil some fresh or dry leaf and may serve directly as green tea. Good for cardiac and diabetic patients. (Smells like Meitei *Mayang Lomba* used in culinary).

24.	*Cynodon dactylon* (L) Pers. [Poaceae] Bermuda/ Durva grass	*Phaiphong* [H]	*Tingthou* [S]	Leaf [O]	Typhoid – Some amount of the tender leaves of *Cynodon dactylon* is crushed well along with some pomegranate. The liquid collected in half a glass is mix with a tea spoonful of honey and is given three times a day before food.
25.	*Desmodium canadense* (Linn.) DC. [Fabaceae] Showy/Canada tick-trefoil	*Kakching threilou* [H]	.... [S]	Leaf [E]	Irregular Menstrual Problem – Boil the leave and serve as decoction a glassful twice a day after food till cured. (The leaves can be served as tea).
26.	*Desmodium rotundifolium.* [Fabaceae], Ticktrefoil	*Mualou* [H]	.... [P]	Whole plant [E]	Blood purifier (esp. for Post-natal women) - Boil and serve as decoction a glassful three times a day before food. One may also have a bath with the extracted boil water. It even helps those regular menstrual for girls.
27.	*Eclipta alba* (Linn.) Hassk. [Asteraceae] Trailing eclipta.	*Uchi sumban* [H]	*Uchi sumban* [S]	Whole plant [E]	Fever (high), Cough – Crush the leaves and the extracted liquid is mixed with little honey and given for cough and fever two tea spoonfuls twice a day.
28.	*Elsholtzia blanda* Benth. [Lamiaceae] Lengser Rep	*Tumpina/ Teckta* [H]	*Lomba* [S]	Leaf [E]	Boil – Heat up the fresh leaves and apply on the boil part with little opening at the boil mouth. Repeat till it is cured.
29.	*Eupatorium odoratum* [Asteraceae], Siam/ Devil weed	*Muitem/ Loukhamong.* [H]	*Chelmen napi* [S]	Leaf [E]	Cuts and Scratches - Boil and have as decoction three times a day a glassful after each food. Crush the leaves and apply on the cuts and scratches to stop blood oozing. It acts as antiseptic.

30.	*Ficus auriculata* (Lour.) Roxb. [Moraceae] Fig tree	*Khrut hei* [T]	*Heirit* [P]	Leaf [E]	Ringworms - Scratch the ringworm area with the fig leaf slightly at an interval time till cured then apply the ringworm area with the young bamboo shoot peeled cover.
31.	*Garcinia xanthochymus* Hook. [Clusiaceae], Dampel.	*Channahei* (Big leaf variety), [T]	*Heibung-asinba* [S]	Leaf [E]	Body Swelling – Heat certain amount of leaves or fruits and use it as balm on the swelling area. It soothes and helps reduce swelling.
32.	*Gynura cusimba* (D. Don) Moore. [Asteraceae] Hill Gynura	*Louran* [H]	*Terapaibi* [S]	Leaf [O]	Hypertension, high blood pressure, anti-inflammation – One may have fresh leaves as chutney or boil the leaves as culinary and have regularly to reduce the high blood pressure and hypertension.
33.	*Gynura procumbent* (Lour.) Merr. [Asteraceae] Subungai	*Tonglaibu* [H]	*Ukhajing* [P]	Leaf [O]	Diabetes - Boiled for about 15-20 minutes some amount of young tender leaves of *Ukhajing* in about three litres of water and served as decoction thrice a day after each meal. The leaves can be eaten fresh along with chutney.
34.	*Kalanchoe K. farinacea* Balf.f [Crassulaceae] Flaming Katy	*Manna-houbi.* [H]	*Manna-houbi* [P]	Leaf [E]	High fever – Crush the leaves and apply the paste on the forehead. Do not eat. The use is for external use only especially for adults.
35.	*Lilium sp.* [Liliaceae],	*Thrunlou* [H]	*Lin-napi* [P]	Tuber/bulb [E]	Snake bites /Dog bites - Crush well this tuber (onion type) and apply on the bitten spot immediately. Do not mix with water or taste the crushed item. (During medication one should not take fruits. It is for external use only, since it is considered highly poisonous.

36.	*Mangifera indica* L. [Anacardiaceae] Mango	*Nouwa hei* [T]	*Heinou* [P]	Leaf [O]	Diarrhoea/ Dysentery – Boil some mango leaves with some water and drink half a glass 2-3 times a day.
37.	*Melastoma malabathricum* Linn. [Melastomataceae] Matakui (Palau)	*Omba heitrung* [H]	*Yachubi* [S]	Leaf, Stem [O]	Diabetes - Boiled some leaves of *Ombouheitrung* in about two litres of water and served as decoction a glassfull twice a day before each meal till cured.
38.	*Melothria maderaspatana* (L.) Cogn. [Cucurbitaceae] Wild cucumber	*Ram machanghei/ Bemangjam* [C]	*Lamthabi* [S]	Whole plant [O]	Jaundice – Boil the whole fresh creeper plant or dry one in about one litre of water and serve as decoction a glass full before meal once a day.
39.	Musa paradisiaca Linn. [Musaceae] Banana flower	*Mut-thro* [Sh]	*Laphu tharo* [P]	Fruit [O]	Dysentery – Slightly roast the banana fruit in the hot fire or heat it in the hot fire ash and have it as food item for 2-3 days or anytime. It helps in suspension.
40.	*Nicotiana tabacum*. Linn. [Solanaceae] Tobacco.	*Heelaknaa/ Sagonta* [Sh]	*Utonglei* [S]	Leaf [E]	a) Fever, b) Joint-sprain – (a) Heat the leaf on the fire and smash little bit and apply on the forehead or body areas. (b) Crush the leaves and apply the paste on the sprain area by re-dressing it at each interval till cured.
41.	*Oroxylum indicum* (L.) Vent. [Bignoniaceae] Indian trumpet flower	*Thampak/ Khe / Shamba* [T]	*Shamba* [S]	Bark, leaf [E]	Tonsillitis/ Sore Throat/ Sinus - Mixed the bark with salt (spring salt) + *tekta* (*lomba*) and boil it and served as decoction half a glass twice a day for a week. The young bean is also edible by roasting.

42.	*Phlogacanthus thyrsiflorus* Nees. [Acanthaceae] Basak	*Simlim/ Ramsingrim* [Sh]	*Nongmangkha Angangba* [P]	Leaf [O]	Cough, Dry cough, Congestion of throat - Crush the leaves and mix with little honey and apply around the chest and inner earlobes and on the head. One may also boil some leaves and have as decoction a glassful twice a day after food till relief.
43.	*Phyllanthus emblica* (L.) [Labiateae] Gooseberry	*Puklu-hei* [T]	*Heigru* [S]	Fruit [O] Bark, Root, Fruit [O]	a) Dry Cough/ Asthma, b) Headache, Hypertension, c) Sore Eyes – a) Crush some amount of goose berry and mix with little honey and have before and after food. b) Eat plenty of gooseberries if one gets headache or feels giddy and drink plenty of water. c) Boil the gooseberry bark or root and splash the liquid on the eyes repeatedly while washing the face in the early morning, afternoon and bed time till cured. Or apply few drops of extracted gooseberry juice directly on the sore eyes 2-3 times a day till cured. Believed to cure even cataract.
44.	*Plumeria acuminate* Ait. [Apocynaceae] Temple/ Pagoda tree	*Leihou khongsang* [Sh]	*Khage-leihou angouba* [P]	Bark [O]	Malaria/ Fever – Crush the dry bark is into powder form and mix half a tea spoon in half a glass of water and take twice a day before each meal. The result is that the stool/ faeces will become black.

45.	*Plumeria rubra* Linn. [Apocynaceae] Red Temple/ Pagoda tree	*Leihou-khongsang Anganba* [Sh]	*Khage-leihou Angangba* [P]	Fruit [O]	Tonsillitis, Malaria – Dry the fruit/seed in the sun and when it turned black, cut into small pieces and eat a piece at any time of the day. Store the small pieces in a tight container. The plant seldom bears such fruit/ seeds. (Remember, do not cut the fruit on any Saturdays).
46.	*Psidium guajava* Linn. [Myrtaceae]- Guava	*Pungtol* [T]	*Pongtol (mana)* [S]	Fruit [O]	Dysentery – Eat the tender leaf and unripe fruits a lot with little salt till relief.
47.	*Punica granalum* Linn. [Onagraceae] Pomegranate	*Kapo-hei* [Sh]	*Ka-phoi* [S]	Fruit, Leaf [O]	Dysentery/ Diarrhoea – Boil some leaves and have as decoction a glass full 2-3 times a day. It helps in suspension.
48.	*Rauvolfia serpentine* Linn. [Apocynaceae], Indian snakeroot.	*Lappar marao* (Red flower). [H]	*Yenbum manbi* [P]	Whole plant, Root [E]	Epilepsy/ Asthma - The roots or whole plant are boiled and served as decoction twice a day after food to cure epilepsy and asthma.
49.	*Rhus semialata* Murr. [Anacardiaceae] Wild varnish tree	*Khongma* [Sh]	*Heimang* [S]	Fruits, Leaf [O]	Diarrhoea, Dysentery, Indigestion – Soaked some of the fruits in a glass of water by adding little common salt and served it to helps indigestion, diarrhoea, and dysentery. Chew some of the tender leave to fight against diarrhoea.
50.	*Saccharum officinarum* L. [Poaceae] Red sugarcane	*Masu marao/ Chu-ngang* [Sh]	*Chu-ngang* [S]	Stem, fruit [O]	Jaundice (*Thongngak*) – Take fresh (red) sugarcane juice daily anytime of the day till it is cured. (Besides for children- Apply the mustard oil on the child's body and exposed to heat and light. Give sugarcane juice daily. Do not give *Ngari* the local fermented dry fish, meat, and chilly, but simple boiled papaya curry).

51.	*Schefflera arboricola* (Hayata) Kanehira. [Araliaceae] Dwarf umbrella plant.	*Lam-om* [Cl]	*Utangbi.* [P]	Leaf [E]	a) Sprain/ Fracture of hands and legs, b) Infertile women – a) Sprain/ Fracture of hands and legs, b) Infertile women. a) Boil with little common salt some leaves of it and make paste by adding little honey and apply on the sprain or fracture part of the leg or hand. b) It is even use in ritual for infertile women.
52.	*Smilax ovaliffolia* Roxb. [Liliaceae] Kumarika/ Aushbah	*Truntu/ Kangra* [Cl]	*Kharam manal* *Kwa-manbi* [P]	Leaf [O]	Urinary/ Kidney/ Stone problems. Boil the leaves in certain amount of water with Sitamasi (white sugar crystal) and serve as decoction a glassful three times a day till relief or cured. The body will swell up a bit and drains out any negative element of body in the form of extracted liquid or urine. (One should not take pumpkin, yam and *ngari* (local fermented dry fish) during the medication).
53.	*Solamum virginuanum* Linn. [Solanaceae] Yellow berried nightshade	*Samtrok-kha* [Sh]	*Leibungkhang* [S]	Fruit [O]	Headache/ Toothache – Crush some amount of the *Solamum* and mixed with two spoonful of honey and have it till the pain is gone.
54.	*Syzigium fruticosum* DC. [Myrtaceae]	*Heinou-manbi* [T]	*Heinou man /* *Tomba-heinou* [S]	Leaf [E]	Fever (especially for children) - Some leaves are boiled and is taken bath with it. (One should not drink the boiled leaves water. For external use only).
55.	*Tamarindus indicus* (L.) [Caesalpiniaceae] Tamarind	*Mangke* [T]	*Mangge* [S]	Seed [E]	Bee Sting – Cut the tamarind seed into half and apply the white part on the bee stung area and bandage it.

56.	*Urena lobata* Linn. [Malvaceae] Ceasarweed	*Chumphal* [H]	*Asikhangla* [P]	Leaf, Stem [O]	Urinary/ Kidney and Gall bladder stone cases - Boil the leaves and stems and have as decoction a glassful three times a day before food till cured.
57.	*Vitex negundo* Linn. [Verbenaceae] Chaste tree	*Warek-lou (Urik-sibi)* [H]	*Urik-sibi* [S]	Leaf [E]	Piles – Burn the dry leaves and collect the ashes and apply on one's anus 3-4 times a day. Or boil some fresh leaves and serve as decoction a glass full before each meal. (Do not eat red meat, chicken and egg during medication. Result is known within 3/4 days).
58.	*Xylos malongifolia* Clos. [Flacourtiaceae] Dandal	*Nungshang-panbi* [T]	*Nunglei shang* [S]	Leaf [O]	Piles – Boil some quantity of the leaves and have 3-4 tea spoonfuls twice a day till it is cured. (The tree has small spikes and the leaves are reddish and shiny when young).
59.	*Zanthoxylum acanthopodium* [Rutaceae] - Pricky winged leaf/ Darmar/ Sheguan pepper	*Singdi* [Sh]	*Mukthrubi* [S]	Seed [O]	Gas Formation - The seeds are crush into powdered form and are mix in half a glassful of warm water and take for 2-3 days, or one may take some along with food items too.
60.	*Zingiber spectabile* [Zingerberaceae], Thai Ruby Ginger (Wild ginger)	*Lamsing* [H]	*Lamsing* [P]	Rhizome [O]	Gas formation, flatulent, stomach upset - Crush the rhizome and add little honey and have as syrup two tea spoonfuls thrice a day till cure.

Abbreviation: *Herb – H, Shrub – S, Tree – T, Creeper – C, Climber – Cl;*

Seasonal - (S), Perennial - (P); Oral - (O), External - (E).

Notes*: Ginseng (Ramhui) is two types: male - female plants. The male species grows with only one stalk while the female plant species has two or more stalks or branches. The male species is considered much more valuable for medicinal purposes.*

Table 2: Distribution of Plant's Family

Sl. No. A	Plant Family	No. of Frequency	Sl.No	Plant Family	No. of Frequency
1.	Acanthaceae	2	20.	Lamiaceae	1
2.	Algae	1	21.	Lauraceae	2
3.	Anacardiaceae	2	22.	Leguminoceae	1
4.	Apiaceae	1	23.	Lilium	2
5.	Apocynaceae	3	24.	Malvaceae	1
6.	Araceae	2	25.	Melastomataceae	1
7.	Araliaceae	1	26.	Meliaceae	1
8.	Asteraceae	6	27.	Moraceae	1
9.	Bignoniaceae	1	28.	Musaceae	1
10.	Caricaceae	1	29.	Myrtaceae	2
11.	Caesalpiniaceae	1	30.	Onagraceae	1
12.	Chenopodiaceae	1	31.	Palmaceae	1
13.	Clusiaceae	1	32.	Phyllanthaceae	1
14.	Crassulaceae	1	33.	Poaceae	3
15.	Cucurbitaceae	2	34.	Rutaceae	1
16.	Fabaceae	2	35.	Solanaceae	2
17.	Flacourtiaceae	1	36.	Verbenaceae	1
18.	Gesneriaceae	1	37.	Zingiberaceae	6
19.	Labiateae	1		Total	60

Table 3: Distribution of Different Types of Ailments and Diseases from the Identified Ethno-botanical Plants

Sl No.	Ailments / Diseases	Frequency	Sl No.	Ailments / Diseases	Frequency
1.	Fever	6	24.	Vomiting of blood	1
2.	Cough	5	25.	Pigmentation/ black spot/Pimples	1
3.	Diabetes	3	26.	Sinus/ sinusitis	2
4.	Burn	1	27.	Typhoid	1
5.	Cut/ Wounds	3	28.	Boil	1
6.	Urinary/ Kidney stone cases	3	29.	Ringworm	1
7.	Dog bites	2	30.	Body swelling	1
8.	Bee sting	2	31.	Anti-inflammation	1
9.	Gas formation / Flatulence	3	32.	Snake bites	1

10.	Piles	3	33.	Diarrhoea	3
11.	Arthritis/ Rheumatism	1	34.	Dysentery	5
12.	Irregular menstrual cycle problem	3	35.	Jaundice	2
13.	Malaria	3	36.	Joint pain	1
14.	Bear/ tiger bites	1	37.	Asthma	2
15.	Stomach burning sensation	1	38.	Headache	2
16.	Ulcer/ Chronic Ulcer	1	39.	Sore eyes	1
17.	Sore throat/ Tonsillitis	4	40.	Epilepsy	1
18.	Hypertension/ High blood pressure	3	41.	Indigestion	1
19.	Worms in babies	1	42.	Sprain/ Fracture	1
20.	Blood purifier/ Post Natal woman	2	43.	Infertility of woman	1
21.	Anaemia	1	44.	Toothache	1
22.	Spike on heel	1	45.	-	-
23.	Child indigestion	1	46.	-	-

Table 4: Distribution of Frequency of Identified Plant's Forms, Types, Parts Used and Forms of Usages

A.	**Forms of Plants**	**Frequency**
1	Herb [H]	26
2	Shrub [Sh]	16
3	Tree [T]	12
4	Climber [Cl]	3
5	Creeper [C]	2
6	Algae	1
	Total	60
B.	Types of Plants	Frequency
1	Seasonal [S]	42
2	Perennial [P]	18
	Total	60
C.	Plant Parts Used	Frequency
1	Leaf	32

2	Bark	6
3	Fruit	11
4	Root	4
5	Whole plant	5
6	Seed	3
7	Stem	5
8	Rhizome	6
9	Tuber/ Bulb	2
	Total	74

D.	Forms of Usages	Frequency
1	Oral [O]	34
2	External [E]	24
3	Both	2
	Total	60

Table 5: The Ethnomedicinal Plants of the Maring Tribe

Sl. No.	**Vernacular Terms**		**Parts Used**	**Ailment and Diseases :**
	Maring Name [H, Sh, T, C, Cl]	**Manipuri Names** [(S)/ (P)]	**[O / E]**	**Treatment Methods and its Dosages -**
1.	*Akdirung* [C]	[S]	Rhizome [O]	Gas formation in stomach – Crush the dry rhizome of this creeper plant in powdered form and mix with little cow's bile juice and have as syrup two tea spoonful twice a day before each food for two-three days. The effective result is indicated by constant excretion of urine and faeces. (They believed that even patient troubled by contact with evil spirits also helps cured).
2.	*Benacha* [C]	--- [S]	Whole plant [O]	Blood Purifier/ Irregular Menstrual/ Urinal Problems - Boil the whole creeper plant (fresh or dry) till the water is deep red and serve as decoction half a glass twice a day regularly before each food for a week. It helps in rejuvenating the internal body organs especially for post-natal women.

3.	*Burmanu* [T]	[S]	Leaf [E]	a) Headache/ Migrants, Giddiness, b) Hypertension - a) Crush the leaves and apply on the top of the head or forehead. One may also mixed with little rice starch for soothing effect. Repeat this process 2-3 times a day till relief. b) Boil some leaves and take bath with the extract water for a week regularly. This is especially for women suffering from hypertension and strong migrants/ headache.
4.	*Chinghaya mathon* [Sh]	[S]	Leaf [E]	Chest pain - Crush the leaves and apply the paste on the chest and other body parts till relief by repeating it at each interval. The plant's fruit is like pea wildly grown in the roadsides and wasteland areas. It is not eatable.
5.	*Hatia* [T]	[S]	Bark [O]	Urinary/ Kidney stone cases - Boil the dry barks in certain amount of water and have as decoction a glassful thrice a day regularly for two weeks.
6.	*Kaksilou* [Sh]	[S]	Leaf [O]	Urinal problem/ Kidney and Gall bladder Stone cases - Boil some leaves till deep red and serve as decoction a glassful thrice a day till relief.
7.	*Kaporlou* [Cl]	*Kang panbi* [S]	Leaf [E]	a) Body swelling (Internal boils), b) Allergy - a) Crush the fresh leaves and apply on the body swelling area for a night. Repeat this process for three or four days till cured. b) Boil the leaves and serve as syrup three tea spoonfuls thrice a day for a week.
8.	*Khamting* [H]	[S]	Root/ Rhizome [E]	Measles, Chicken pox – Crush with little water and add little bear's bile juice and apply on the whole body. This is especially for children suffering from measles and chicken pox.

9.	*Khanglung* [Cl]	[P]	Leaf [E]	Sprain/ Fracture of hands and legs - Crush the leaves and apply the paste on the sprain or fracture parts by wrapping it for a day or two.
10.	*Kilnambo / Sanarei* [H]	[S]	Leaf [O]	High fever - Boil the leaves and have three times a day as decoction half a glass till cured.
11.	*Kinneem* [Sh]	*Kongpe mana / Kinneem* [S]	Leaf [O]	Fever - Boil some leaves of *Kinneem* in a liter of water and drink a glassfull twice a day after food till the fever is reduced. (The leafs looks like Jumun leafs).
12.	*Kongkohui* [H]	[S]	Roots, Leaf [O]	a) Diabetes, b) Urinary/ Kidney stone cases - a) Boil some roots of this shrub plant with some water and have as syrup two tea spoonfuls three times a day till cured. b) Boil the leaves with some Sitamasi sugar (white sugar cube) and have a syrup three teaspoonful three times a day. The urine excreted will be black in colour.
13.	*Kuldang* [Sh]	[P]	Leaf [E]	Snake bites - Crush the male plant leaves of this shrub plant and apply the paste in bitten area.
14.	*Kulbirui* [Cl]	*Wangbarel* [P]	Bark, stem [E]	Heart and Kidney Failure Problems - Boiled half a Kilogram of the climber bark in about 5 litres of water till it turns deep red. Bath with the extracted boiled water or apply regularly on the chest and abdomen pain areas any time of the day regularly for a week or more. This medication is exclusively for external purposes only. Therefore, do not drink the extracted liquid.
15.	*Kursi* [C]	[S]	Fruit [E]	Cuts, wounds, sprain - Crush the fruits of this creeper plant into powder form and apply on the cuts and wounds areas till cured by changing the bandage.

16.	*Lamkumlou* [H]	*Chakmang mana* [S]	Roots, Leaf [O]	Female blood purifier (Post-natal woman) - Boil and have two tea spoonfuls as syrup three times a day of this herb plant.
17.	*Loumeh* [H]	[S]	Root [O]	Diarrhoea, Dysentery, Typhoid - Boil the roots of this herb plant till deep red and serve as decoction a glassfull three times a day till cured. The dry root also can be used. (Do not take chilly, fruits, prawn, eel during this medication).
18.	*Loulurpui* [H]	*Hameng sampakpi* [S]	Leaf [O]	Typhoid - Boil the leaves of this herb plant and have as decoction half a glass twice a day till cured.
19.	*Louthaither* [H]	[S]	Root [O]	High fever/ tooth ache - Boil the roots and have as syrup two spoonfuls twice a day till cured.
20.	*Lungpaar* [H]	*Nungthambal* [S]	Stem/ Flower [E]	Cancer - Smear the stem/ flower on a clean hard stone by mixing with little water and collect the extracted liquid with cotton and apply it on the cancer part and bandage it well. Repeat this formula after two days and continue for 3-4 months. It should be done before chemotherapy treatment. This is for external used only. Other items and ingredients are also used. (It is like rock mushroom).
21.	*Ngalpho* [H]	[S]	Leaf [E]	Fever - Crush the leaves and apply the paste on the forehead and other body parts to reduce the fever.
22.	*Nunghoak Sirik* [H]	[S]	Root [E]	Leprosy - Extracted oil of the roots is applied on the whole body parts time to time till cured.

23.	*Pankhok* [H]	[P]	Rhizome [E]	Body swelling - Crushed the rhizome and add the extracted liquid of the thunder-bolt stone and apply on the swelling body parts. This is a kind of wild water yam.
24.	*Rultumlou* [C]	[S]	Whole plant [O]	Irregular menstrual problems, Blood purifier (Post-natal woman) - Boil the whole creeper plant and serve as decoction a glassful by mixing a spoonful of honey in a glass and have three times a day for a week. The boiled extracted liquid is red in colour. This herbal medicine is especially recommended for post-natal women who suffered from irregular menstrual or for irregularity.
25.	*Sakon hidak trung/ Mickyekpaar* [Sh]	*Kothap mana* [S]	Leaf [E]	Removal of bad blood, poisonous insect bites, boil - Crush the leaves and apply the paste on the swelling area of the bad blood, or bite of any poisonous insect bites or boil area till cure.
26.	*Salda wa* [H]	*Nat wa/va* [P]	Leaf [O]	Tonsillitis - Boil some of this bamboo leaves and have as decoction three times a day till cured the tonsillitis. During recovery one would feel thirsty for water.
27.	*Taangkha/ Taangkinu* [H]	*Yanungkha* [S]	Whole plant [O]	Tonsillitis/ Sore Throat Problem - Boil the leaves or some amount of this *Yanungkha* whole plant in about three litres of water and have as decoction a glassful twice a day after each meal. One may also have it in raw fresh form till it helps cure. (Do not give to girls who are weak and suffering from leukaemia). Some claimed it helps fight against cancer, boil types, stomach ailments too.

28.	*Thingkangphu* [T]	[P]	Bark, Leaf [O]	a) Liver problem (alcoholic), Ulcer, b) Gastritis and Stomach problems – a) Boil the bark of this tree plant with some amount of water and serve as decoction twice a day for a week to help relief from chronic ulcer, liver and stomach problems. This is especially recommended for patient suffering from alcoholism. b) Boil the leaves in about three litres of water and served as decoction half a glassful twice a day before food till it cures. (Do not eat spicy things like chilly, ginger, etc. during medication).
29.	*Thingphungcho* [Sh]	[S]	Leaf/ Bark [O]	Urinary/ Kidney/ Gall bladder Stone Cases - Boil the leaves in about five litres of water and served as decoction a glassful thrice a day before food till it is cures. This species is believed to serve as anti-oxidant too.
30.	*Thruna* [H]	[S]	Leaf [O]	Urinary / Kidney Stone problems - Boil some amount of the leaves till deep red and drink as decoction three times a day till relief of pain. The urine will flow freely.
31.	*Tumpina* [H]	[S]	Leaf [E]	Boil - Crush the leaves and apply the paste on the boil area by repeating at certain intervals.
32.	*Waa-aryam* [Cl]	[P]	Leaf [E]	Body-ache, Weakness, Post-natal - Take bath with the boiled fresh leaves and one may used as bam with the hot extract water regular till relief.
33.	*Yaili* [Cl]	*Ngamu yai* [P]	Root [E]	Scabies - Crush the roots of this plant and apply the paste on the scabies area three-four times a day till cured. The crush roots and barks are also used for fishing in the river cleavages by locals.

34.	*Yasikhong* [Cl]	[P]	Rhizome/ Root [E]	Snake bites, dog bites - Crush the rhizome of this Yasikhong climber and apply the paste on the bitten part till cured.
35.	*Youlou* [H]	[S]	Root [O]	Urinary / Kidney Stone Problems - Boil some leaves with Sitamasi (White sugar free cube) and serve as deoction a glassful three times a day till relief. It is colourless and has no smell. (It is small herbal plant with small leaves and small white flowers).

Table 6: Distribution of Frequency of Different Types of Ailments and Diseases of Maring

Sl No.	Ailments / Diseases	Frequency	Sl No.	Ailments / Diseases	Frequency
1.	Fever	4	16.	Snake bites	2
2.	Diabetes	1	17.	Diarrhoea	1
3.	Cut/ Wounds	1	18.	Dysentery	1
4.	Urinary/ Kidney stone cases	7	19.	Allergy	1
5.	Dog bites	1	20.	Joint pain/ Body ache/ Weakness	1
6.	Gas formation / Flatulence	1	21.	Chest pain	1
7.	Irregular menstrual cycle problem	2	22.	Headache	1
8.	Liver / Ulcer problems	1	23.	Leprosy	1
9.	Sore throat/ Tonsillitis	2	24.	Removal of bad blood/ Poisonous insect's bites	1
10.	Hypertension/ High blood pressure	1	25.	Sprain/ Fracture	1
11.	Blood purifier/ Post-Natal woman	3	26.	Heart/ Kidney failure	1
12.	Typhoid	2	27.	Toothache	1
13.	Boil	2	28.	Cancer	1
14.	Body swelling	1	29.	Measles	1
15.	Scabies	1	30.	Chicken pox	1

Table 7: Frequency Distribution of Ethnomedicinal Plant's Forms, Types, Parts Used and Forms of Usages

A.	**Forms of Plants**	**Frequency**
1	Herb [H]	16
2	Shrub [Sh]	6
3	Tree [T]	3
4	Climber [Cl]	6
5	Creeper [C]	4
	Total	35

B.	**Types of Plants**	**Frequency**
1	Seasonal [S]	26
2	Perennial [P]	9
	Total	35

C.	Plant Parts Used	Frequency
1	Leaf	19
2	Bark	4
3	Fruit	1
4	Root	9
5	Whole plant	3
6	Stem	2
7	Rhizome	4
8	Flower	1
	Total	43

D.	Forms of Usages	Frequency
1	Oral [O]	18
2	External [E]	17
	Total	35

Notes*: The total number is more because some plant's part of the same plant is used either for the same purpose or differently like leaf/ bark, root/ leaf, etc.*

Table 8: Category Wise Distribution of Total Frequency of Ethno-botanical Plants of Maring

Sl No.	**Types of Category**	**Chapter 2**	**Chapter 3**	**Total**
1	Total number of Identified Ethno-botanical plants species	60	-	**60**
2	Total number of Ethnomedicinal plants	-	35	**35**
	Total	60	35	95
	Total number of Plant's Family	37		37

Table 9: Distributions of Total Frequency of Common Types of Ailments and Diseases among Maring (and Khoibu) Tribe of Manipur

Sl No.	Ailments / Diseases	Frequency Table 3	Frequency Table 6	Total	Sl No.	Ailments / Diseases	Frequency Table 3	Frequency Table 6	Total
1.	Fever	6	4	**10**	28.	Boil	1	2	3
2.	Cough	5	-	**5**	29.	Ringworm	1	-	**1**
3.	Diabetes	3	1	**4**	30.	Body swelling	1	1	2
4.	Burn	1	-	**1**	31.	Anti-inflammation	1	-	**1**
5.	Cut/ Wounds	3	1	**4**	32.	Snake bites	1	2	3
6.	Urinary/ Kidney/ Gall bladder Stone cases	3	7	**10**	33.	Diarrhoea	3	1	**4**
7.	Dog bites	2	1	**3**	34.	Dysentery	5	1	6
8.	Bee sting	2	-	**2**	35.	Jaundice	2	-	**2**
9.	Gas formation / Flatulence	3	1	**4**	36.	Joint pain/ Body ache/ Weakness	1	1	2
10.	Piles	3	-	**3**	37.	Asthma	2	-	**2**
11.	Arthritis/ Rheumatism	1	-	**1**	38.	Headache	2	1	3
12.	Irregular menstrual cycle problem	3	2	**5**	39.	Sore eyes	1	-	**1**
13.	Malaria	3	-	**3**	40.	Epilepsy	1	-	1
14.	Bear/ Tiger bites	1	-	**1**	41.	Indigestion	1	-	**1**
15.	Stomach burning sensation	1	-	**1**	42.	Sprain/ Fracture	1	1	2

16.	Ulcer/ Chronic Ulcer	1	1	**2**	43.	Infertility of woman	1	-	**1**
17.	Sore throat/ Tonsillitis	4	2	**6**	44.	Toothache	1	1	2
18.	Hypertension/ High blood pressure	3	1	**4**	45.	Heart/ Kidney failure	-	1	**1**
19.	Worms in babies	1	-	**1**	46.	Cancer	-	1	1
20.	Blood purifier/ Post Natal woman	2	3	**5**	47.	Removal of bad blood/ Poisonous insect's bites	-	1	**1**
21.	Anaemia	1	-	**1**	48.	Leprosy	-	1	1
22.	Spike on heel	1	-	**1**	49.	Chest pain	-	1	**1**
23.	Child indigestion	1	-	**1**	50.	Allergy	-	1	1
24.	Vomiting of blood	1	-	**1**	51.	Scabies	-	1	**1**
25.	Pigmentation/ black spot/ Pimples	1	-	**1**	52.	Measles	-	1	1
26.	Sinus/ sinusitis	2	-	**2**	53.	Chicken pox	-	1	**1**
27.	Typhoid	1	2	**3**		**Grand Total**	85	45	130

Table 10: Category Wise Distribution of Total Frequency of Plant's Forms, Types, Parts Used and Usages

A.	Forms of Plants	Frequency Table 4	Frequency Table 7	Total
1	Herb [H]	26	16	**42**
2	Shrub [Sh]	16	6	**22**
3	Tree [T]	12	3	15
4	Climber [Cl]	3	6	**9**
5	Creeper [C]	2	4	**6**
6	Algae	1	-	**1**
	Total	60	35	95

B.	Types of Plants	Frequency	Frequency	Total
1	Seasonal [S]	42	26	**68**
2	Perennial [P]	18	9	**27**
	Total	60	35	95

C.	Plant Parts Used	Frequency	Frequency	Total
1	Leaf	32	19	**51**
2	Bark	6	4	**10**
3	Fruit	11	1	**12**
4	Root	4	9	**13**
5	Whole plant	5	3	**8**
6	Seed	3	-	**3**
7	Stem	5	2	**7**
8	Rhizome	6	4	**10**
9	Tuber/ Bulb	2	-	**2**
10	Flower	-	1	**1**
	Total	74	43	117

D.	Forms of Usages	Frequency	Frequency	Total
1	Oral [O]	34	18	**52**
2	External [E]	24	17	**41**
3	Both	2	-	**2**
	Total	60	35	95

Table 11: Other Forms of Indigenous Medicines Used by Maring Tribe

Sl No	Common Term	Maring Name	Meitei Name	Parts Used	Diseases
1	Crystal cane sugar	*Sitamasi*	*Sitamasi/ Misri* in Assamese	Sugar liquid	Some amount of this crystal cane sugar is added as an ingredient in the preparation of decoction and syrup like in *Centella asiatica* liquid extract, or while boiling the indigenous herbs or barks or creeper or climber especially, to enhance the taste from extreme bitterness or clumsy taste.
2	Raw rice	*Cheng*	*Cheng*	Raw Rice	Dog's bite - Chew the raw rice and apply the paste on the dog's bitten part and paste a silver coin on it.
3	Fermented local rice beer	*Khaji (chakthamwa)*	*Zumapang*	Fermented rice beer	Urinal/ Kidney Stone Problem – Have 3-4 mugs of sticky fermented rice beer in a day for 2-3 months.
4	*Warhog / Kitri* (An unique animal with red and bluish slicken plume that looks like a wild boar and has a limb/ hoof like a dog).	*Lam-ok*	*Lam-ok manbi*	Blood	Leprosy (Richi in Maring) – The person infected with leprosy may apply the blood of this unique animal regularly and also eat the cooked meat. It is believe that it cures within a weeks' time.
5	*Lethocerus americanus* [Belostomatidae] Giant Water Bug	*Nausek* (Lake's species)	*Nausek*	Head	Snake Bite/ Bitten by any Poisonous Reptiles - Get the head of the lake's *nausek* insect species and stuck its head at the bitten part. It stuck and sucks out all the venoms automatically till it is clear. The *nausek* head will drop once it finishes sucking it.

6	*Planorbellatrivolvis* [Planorbella] Gastropod mollusk (Freshwater snail)	*Tharoi*	*Tharoi*	Membrane	Snake Bite – Get a fresh snail and put the head around the bitten part without disturbing it for a while. It is believe to suck the venoms out.
7	Dried bile juice/ Any dried animals' bile juice	*Mit (lit. bile juice)*	*Masingkha*	Bile juice	Measles, Chicken pox. – Python's bile, Beer's bile, Black monkey's bile, and crow's bile all these bile mixed together with little water is applied on the body to help cure the measles and even chicken pox. (Most of the animals' bile juice dry or fresh is considered very usefull for local medicine man in the preparation of their indigenous medicines for the treatment of various diseases).
8	Thunderbolt stone	*Thaangmuna rei*	*Braja nung*	Extract liquid of thunder bolt stone	Fever, Stomach upset The thunderbolt stone is rub against the hard stone by adding little water and the extracted liquid is collected with cotton and apply on the forehead, body and is also suggested to drink little to cure the fever and stomach upset when believe to have caused by evil spirits.
9	*Vespulagermanica* F. [Arthropoda] House wasp	*Leibak Khoi*	*Leibak Khoi*	Mud	Mumps (Type) – Mixed the mud collected from house wasp with saliva or with little water and apply it on the mumps area. Apply 3-4 times a day. (It cures but reason unknown).

Table 12 :Preparation Methods of Some Compound Ethnomedicines Used by Maring Tribe

Sl. No	Diseases	Maring Name	Manipuri Name	Botanical Name	Parts [Forms - (O)/ (E)]	Treatment Methods and its dosages
1.	Typhoid –	*Phaiphong*+ *Kapo-hei* + *Khoihi* (Honey)	*Tingthou* + *Kaphoi* + Honey	*Cynodon dactylon*+ *Punicagranalum* + Honey	Leaf [O]	Some amount of the tender leaves of *Cynodon dactylon* is crush well along with some pomegranate. The liquid collected in half a glass is mix with a tea spoonful of honey and given three times a day before food.
2.	Dry Cough/ Throat Congestion -	*Ramsinrim* + *Triptung-ngou* + *Singdi* + *Hui* + (white Sugar crystal)	*Nongmangkha-anganba* + *Nongmankha-angouba* + *Mukthrubi* + *Sing* + *Sitamasi*	*Phlogacanthus thyrsiformis Nees*+ *Zanthoxylum acanthopodium*+ Ginger + white sugar cube	Tender leaf + tender leaf + seed + rhizome + sugar	Some amount of the leaves, seeds and tuber along with white sugar cube are boil together till the colour turns blackish and half a glass full is serve as decoction before food. (During medication one should not take red meat or *ngari* (fermented dry fish).
3.	Ringworm –	*Heirik mana* + *Souhing-khor*	*Heirik mana* + *U-soi*	Ficus cunia (Moraceae)+ *Schima wallichii* Choisy	Leaf + stem knot	Scratch the ringworm area with the fig leaf then, apply the hairy stem knot on it.

4.	Muscle Swelling	*Pankhok-moinum* + *Shamlolubal* + *Umkhabal* + *Thunder bolt stone*	*Pankhokmoinum* + __ +___ +*Tinkhangnung*	A type of yam + __ + __ + thunder bolt stone	Leaf + root + root + liquid	Crush the leaves and roots together and mixed the thunder bolt stone liquid collected from rubbing against a hard stone and apply the paste on the spot time to time till cures.
5.	Fever (Malaria Type)	*Kylnum* + *Sanarei* + *Chilly* + *Taa-nataret* bamboo straps	__ + Sanarei + Marok + Taa-nataret	___ + Marigold + Chilly + Seven straps of bamboo	Root + leaf + fruit + stem	Mix well all items and wrap it in a cloth and tie around the ankle with seven small bamboo straps.
6.	Worm/ (Ascaris)	*Puleimanbi* + *Kani*/ or kerosene	Pullei + Kanii or kerosene	*Alpinia allughas* + product of tobacco or kerosene	Root + solid or liquid	The mixture is apply on the neck, throat and chest esp. for babies.
7.	Measles (*Leikup* in Manipuri)	Python's bile + bear's bile + black monkey's bile + crow's bile	Leiren + sawom + Jong + Kuwak*mashingkha*	Leiren + Sawom+ Jong +Kuwakmashingkha	Python's + Bear's + Monkey's +Crow's bile juices	Mixture of all these bile juice is given in small dosage half a tea spoonful to children and a tea spoonful for an adult after food twice a day.

CHAPTER - 5

◆◆◆◆◆

Conclusion

The study on indigenous medicine among the Maring (and Khoibu) tribe is of first kind among them. The survey is based on randomly selected villages of Maring (and Khoibu) inhabiting in Chandel, Tengnoupal and Senapati districts of Manipur. The selected respondents are village chiefs, priests, priestess, village elders, and senior citizens and leaders who are knowledgeable in the practice of ethnomedicines. Some villages are located deep in the interior near Myanmar, some on the top of the mountains, while others on the foothills near the valley.

The book is divided into five chapters. The Chapters One introduces about the importance of ethnomedicines and how the indigenous traditional value system is declining fast with the changing world even among the Maring tribe of Manipur. The Chapter Two describes the various identified Ethno-botanical plants with photographs used as medicines by Maring (and Khoibu) tribe of Manipur in the treatment of different ailments and diseases. The Chapter Three also describes other significant ethnomedicinal plants provided in vernacular terms and their usages besides, other forms of indigenous medicines used by Maring tribe comprising of insects, animals, and other valuable items. The Chapter Four is the analysis of the data, while Chapter Five is the discussion that culminates as part of conclusion.

There are a total of 95 (Ninety five) ethno-botanical plants discussed in the book besides, 9 (nine) other important forms of indigenous medicines. There are a total of 12 Tables arranged based on the analysis according to the chapters. The Table 8 shows that out of 95 ethno-botanical plants 60 (sixty) species are identified belonging to 37 (thirty seven) plant families, while the other 35 ethnomedinal plants are yet to be identified.

The Tables 1 & 5 describes briefly the various identified and non-identified ethno-botanical plants of Maring tribe and their usages, thereby providing the Forms of the plants, Types of plants, Forms of Usages and the treatment methods.

The Table 2 indicates that there are 37 plant's families from the identified 60 plant species. The highest frequency are shared by: (i) Asteraceae – 6, (ii) Zingiberaceae – 6, (iii) Apocynaceae – 3, (iv) Poaceae – 3, (v) Acanthaceae – 2, (vi) Anacardiaceae – 2, (vii) Araceae – 2, (viii) Cucurbitaceae – 2, (ix) Fabaceae – 2, (x) Lauraceae – 2, (xi) Lilium – 2, (xii) Myrtaceae – 2, and, (xiii) Solanaceae - 2. The remaining plant families shared one frequency each.

The Asteraceae and Zingiberaceae families are widely used because of the availability and the variety found in the region and for its remedy effectiveness. It may be noted that some species are not exactly the same species mentioned though belong to the same family because of the variation found in the region. Therefore, the pictures provided will help support in identifying the exact species.

In Table 3 different types of ailment and sicknesses indicates that there are 44 (Forty four) sicknesses. The frequencies are: (i) Fever – 6, (ii) Cough – 5, (iii) Dysentery – 5, (iv) Sore-Throat/ Tonsillitis – 4, (v) Diabetes – 3, (vi) Cuts and wounds – 3, (vii) Urinary/ Kidney/ Gall bladder stone cases – 3, (viii) Gas formation/ Gastritis (Flatulence) – 3, (ix) Piles - 3 and (x) Irregular menstrual problems - 3, (xi) Malaria – 3, (xii) Hypertension – 3 and, (xiii) Diarrhoea – 3. The rest of the ailments have a frequency of two and one each. From the above data the most common ailments and diseases are: Fever, Cough, Dysentery and Tonsillitis. However, other ailments cannot be ignored.

Table 4 in the category - (A) *Forms of Plants* from the identified ethno-botanical plants it indicates that herbs, shrubs and trees are most common plants used in healing and curing different ailments and diseases by Maring people. The maximum frequency is shared by herbal plants. The frequencies are: (i) Herb – 26, (ii) Shrub – 16, (iii) Tree – 12, (iv) Climber - 3 and, (v) Creeper – 2 and (vi) Algae -1.

Table 4 in the category (B) *Types of Plants*, it indicates that maximum plants are seasonal (S) based with 42 species while perennial (P) plants shared 18 species.

Table 4 in the category (C) *Parts Used*, the maximum frequency is shared by leaf. The frequencies are: (i) Leaf - 32, (ii) Bark – 6, (iii) Fruit - 11, (iv) Root – 4, (v) Whole plant – 5, (vi) Seed – 3, (vii) Stem – 5, (viii) Rhizome – 6, (ix) Tuber – 2, and (x) Algae - 1.

Table 4 in the category (D) *Forms of Usages* indicates that the Maring used more these ethnomedicines in the Oral (O) form of traditional decoction and syrup. The frequencies are: Oral – 34, the External - 24, and in Both – 2. Often the traditional form of decoction and syrup dosages is taken as a glassful or half a glass twice/ thrice a day or two/three tea spoonful a day till it is cured. The medication for external are prepared in balm or paste forms while some are applied as ointment externally on the infected or injured sprain parts. These plants are considered mostly not suitable for eating or drinking since poisonous.

Table 5 describes another 35 non-identified ethnomedicinal plants of Maring used as medicines. Under this list Table 6 indicates that there are about 30 types of ailments and diseases. Here the highest frequencies are shared by: (i) Urinary/ Kidney/ Gall bladder stone cases – 7, (ii) Fever - 4, (iii) Blood purifier/ Post-Natal woman - 3, (iv) Irregular menstrual problems – 2, (v) Typhoid - 2, (vi) Boil - 2 , (vii) Sore throat/ Tonsillitis – 2 and, (viii) Snake bites - 2. The rest of the ailments shared a frequency of one each. Unlike in the previous list the Table 6 indicates that the most common and vulnerable sicknesses are urinary/ kidney/ gall bladder stone cases, blood purifier/ Post-natal woman, irregular menstrual problems and snake bites. These sicknesses are not new but have become common among them.

These ethnomedicinal plants though not very common in practice are considered very useful medicinal plants in certain respect as it healed and cured their sicknesses. Most of these plants are rare or grown in the deep interior forest and very few people knew the medicinal values. Therefore, it is not so common among them, although some who knew about its effectiveness and values used them at times when pharmaceutical medicines are not available or as a supplement.

In the survey, it is observed that many Marings still depends on their indigenous medicinal plants in the treatment of various ailments and diseases apart from pharmaceutical medicines. The common ethno-botanical plants that are planted around their houses for used are like: *Mukiama deraspatana (Bemanjam), Oroxylum indicum (Shamba), Zanthozylum acanthopodium (Singdi), Solanum virginuamum (Samtrok-kha), Benincasa hispida (Anmahei-angou), Eupatorium birmanicum (Langthrei), Benacha, Thingphungcho,* etc. The plants liked *Bemanjam, Benacha, Thingphungcho* are dried and preserve for future used. In this regard, certain plants are also similarly used by other communities but the application methods in the treatment of ailments are sometimes different.[1]

In continuation of the analysis the Table 7 ethnomedicinal plants of Maring it indicates that in the category of (A) *Forms of Plant* the Maring used maximum herbs compared to other forms of plants. The frequencies are: (i) Herb - 16, (ii) Shrub - 6, (iii) Tree - 3, (iv) Climber – 6 and, (v) Creeper – 4.

Table 7 in the category (B) *Types of Plants,* the frequencies shared are: Seasonal (S) - 26 and Perennial (P) – 9. Herein, it indicates that the Maring used more seasonal plants in their treatment of different ailments which are mostly herbs.

Table 7 in the category (C) *Parts Used,* the highest frequency is leaf indicating the significance of it. The distributions of frequencies are: (i) Leaf - 19, (ii) Bark - 4, (iii) Fruit - 1, (iv) Root – 9, (v) Whole plant – 3, (v) Stem - 2, and (vi) Rhizome – 4 and, (vii) Flower -1.

We also noticed that in the category (D) *Forms of Usages,* the maximum frequency is oral. The frequencies are: Oral (O) – 18 and External (E) – 17.

However, Table 9 from the analysis of Table 3 & 6 of the distribution of *total frequency of Common types of ailments and diseases* there are approximately 53 (Fifty three) different types of ailments and diseases. The Table 9 shows the most vulnerable 19 ailments and diseases noticed among the Maring tribe. They are: (i) Fever – 10, (ii) Urinary/ Kidney/ Gall bladder stones cases – 9, (iii) Tonsillitis – 6, (iv) Dysentery – 6, (v) Blood purifier esp. of post-natal woman – 6, (vi) Irregular menstrual problems – 5, (vii) Cough – 5, (viii) Blood purifier/ Post-natal woman – 5, (ix) Diarrhoea – 4, (x) Gastritis/ Gas formation – 4 and (xi) Diabetes – 4, (xii) Cut and Wounds – 4, (xiii) Typhoid – 3, (xiv) Dog bites – 3, (xv) Malaria – 3, (xvi) Piles – 3, (xvii) Headache – 3, (xviii) Snake bites – 3 and, (xix) Boil – 3. The rest of the ailments and diseases have frequencies of two and one each under the listed table.

Although, Table 9 shows that fever, cough, dysentery, diarrhoea, gastritis, piles, ulcer shows high frequencies these illnesses are very common and not so dangerous. They can be easily treated with these ethnomedicines and shows remarkably effective for some. However, the new deadly diseases like diabetes, urinary/kidney/gall bladder stone cases, irregular menstrual problems of women, cardiac heart problem, malaria and snake bites have become common now amongst the Marings too. These deadly ailments may be seriously thought about since these illnesses are increasing at higher rates among them.

The diabetes sickness, although recently known to them have become common among them too for certain reasons like food habits and living styles. In this context, the local medicine men have been experimenting with different indigenous herbal plants. Some patients have even claimed that it was effective in controlling their blood sugar levels and they felt better while others find not so effective. Therefore, it depends upon the type of patient and the medicine man consulted.

Certain local Maring medicine men and women claimed to heal and cure like cancer, breast cancer, cardiac heart problem, urinary/ kidney stone cases, liver and chronic ulcer problems, jaundice and so on with their ethno-botanical plants. In certain cases, some add other hidden valuable items comprising like animal's objects, insects, and other materials known to them in curing and healing such deadly diseases.

For example, Mr. Menai a popular local medicine man of Maring and a respondent who lives at Pallel said that "I can cure breast cancer and certain cardiac problem completely with my herbal medicines, if the patients strictly followed my prescription properly".

Another significant ailment observed is the irregular menstrual problems. This natural menstrual cycle of woman is considered a serious health concern for every woman that often worries them in their life, irrespective of caste, creed, colour or race, whether she is educated or illiterate, or she lives in city, town or village. When her menstrual cycle is disturbed from normality they knew it is a major health concern. When a Maring woman suffered from such irregular menstrual cycle or wants to flush out bad blood after post-natal or delivery of a child they used certain traditional ethnomedicinal plant's leaf or bark or stem that belong to herbs, shrubs or creepers often in the form of decoction and syrup to cure and heal such problems or to enhance blood purification.

In the survey, it was found that at Kakching Lamkhai bazaar and Pallel bazaar the local Maring vendors sold different dried ethnomedicinal leaves, barks and creepers while some are packed in polythene. These ethno-botanical plants are prescribed by their local medicine man for the use of jaundice, cancer, hypertension, urinary/kidney stone cases, irregular menstrual problems, gastritis/ chronic ulcer, and blood purification for post-natal woman and so on. For example, the *Benacha* is used for irregular menstrual problems, *Thingkangphu* for the treatment of chronic ulcer and gastritis, *Melothria maderaspatana* (L.) Cogn *(Ram machanghei/ Bemangjam/ Lamthabi)* for serious jaundice case, and *Thingpungcho* for urinary/kidney/gall bladder stones cases.

These dried ethnomedicinal plant items are being used in the region especially, by the Maring and neighbouring communities for its effectiveness and in controlling their sicknesses and those who considered helpful continues to purchase from them (see Table 5). The increase rate of urinary/ kidney/gall bladder stone problems and tonsillitis among the Maring suggest that they drank less water or drank contaminated or unboiled/ unfiltered water from any water sources like most hill tribes.

Traditionally, in the past most tribal communities of Northeast India took certain herbal leaves, barks, creepers, or climbers as decoction and sometimes given even to post-natal woman even if she does not suffer from such menstrual cycle problems as precautionary measures, so that, she does not suffer in future and other related health problems, though now they seldom used.

Another important ailment is the snake bites which cannot be ignored as they mostly inhabit in the hilly forest areas and ventures into dense forest for gathering the forest produced. They used the tuber of *Lilium* sp. (*Thrunlou* / Lin-napi), leaf of *Kuldang* and the rhizome of *Yasikhong* when snake bites them. Dog bites is also another common ailment since most Maring families rear dogs at home that often attacked passerby or while visiting the family. So, in case of dog bites they used *Albizia myriophylla* Benth (*Threlou or Yanglee* in Meitei) for dog bites and when not available some used unripe papaya (*Carica papaya* L.) for emergencies.

Mr. M Medun of Sandan Senba said that "Depending upon the need and urgency we sometimes smeared with the stem of the wild yam *Alocasia macrorrhiza* (Linn.) (*Hongngo*) when the bee or hornet stung us". The itchiness seems to reduce the pain.

There are also other respondents who claimed expertise in their own field that they can cure and heal like internal bleeding of intestines or chronic ulcer or piles, if the patients took their prescribed medicine properly and regularly. These medicine men and women claimed of no side-effects using their medicines but they do caution that certain food habits like oil, red meat, chilly and spicy items should be avoided during the medication period.

However, the fees for the treatment are normally expensive to very expensive compared to the home remedies treatment since, the sicknesses are deadly and the treatment involves several stages of application that usually last for a month to three months or more.

Further, the Table 10 (analysis of Table 4 and 7) in distribution of total frequency in the category of - (A) *Forms of Plant*, the herbs and shrubs shared the maximum frequencies in the treatment of various ailments and diseases. The frequencies are: (i) Herbs – 42, (ii) Shrub – 22, (iii) Tree – 15, (iv) Climber – 9, (v) Creeper – 6 and (vi) Algae – 1.

The Table 10 in the category distribution of total frequency of (B) *Types of Plants* there are a total of 68 Seasonal (S) and 27 Perennial (P) plants in all. The frequency indicates that the Maring used maximum seasonal ethno-botanical plants as medicines.

Table 10 in the category of distribution of total frequency of (C) *Parts Used* it indicates that the maximum frequencies are shared by the leaf, root and fruit. The frequencies are: (i) Leaf – 51, (ii) Bark – 10, (iii) Fruit – 12, (iv) Root – 13, (v) Whole plant – 8, (vi) Seed - 3, (vii) Stem – 7, (viii) Rhizome – 10, (ix) Tuber/ Bulb – 2 and, (x) Flower - 1. .

The Table 10 in the category distribution of total frequency of (D) *Forms of Usages*, there are a total of: (i) Oral – 52 and, (ii) External – 41 and, (iii) Both - 2. The frequency indicates that the medication methods are mostly in the traditional form of decoction and syrup, where the external are treated in the forms of balm, ointment and paste, while two species are used in either ways in the application methods.

In the survey, it is found that most of these indigenous medicinal plants belonging to herbs, shrubs, creepers and climbers appeared clean, harmless, less awful or fearful or less dangerous, although some are slightly clumsy and bitter in taste.

In this respect, it is seen that certain plants like *Centella asiatica* Linn (Peruk), *Curcuma caesia* Roxb (Yaimu), *Rhus semialata* Murr (Heimang), *Benincasa hispida* (Khulbi/Torbot), *Oroxylum indicum* (Shamba), *Phyllanthus emblica* (Goose berry), *Antidesma acidum* (Ching-yensil) and etc. are not only highly anti-oxidant but are rich in multi-vitamins and minerals like iron, calcium, potassium, manganese, phosphorus, etc. that helps in speedy recovery of certain ailments. These cannot be denied because most of the plants and fruits grown on the hills and top soil are consider very rich in minerals which served as medicinal properties besides, they are free from any poisonous substances like insecticides or pesticides or artificial fertilisers as they grew naturally. These plants need to be closely examined and analyse scientifically by scientists the constituent components.

The Table 11 shows 9 (nine) other forms of indigenous medicines of the Maring tribe like insects, animals, thunderbolt stone and others that have significant medicinal values among the Maring and other Northeastern tribes of India. They are like: (1) Crystal cane sugar (*Sitamasi* or *Misri* in Assamese), (2) Raw rice, (3) Fermented rice beer, (4) *Kitri* (Warhogs), (5) Giant water bug, (6) Snail, (7) Dried pig's bile, (8) Thunder bolt stone and, (9) House wasp mud. Some of these items are considered highly valuable because of its effectiveness and are preserved in dried form for future used.

These items mentioned above are not very familiar with many people. They are used as single or as compound with other essential items. For examples, the mud of House wasp (*Vespulagermanica* F.) mixed with little saliva or water is applied on the mumps area. They claimed if applied regularly for few days the mumps subsides and is healed. The Giant Water Bug (*Lethocerus americanus*) is considered very effective in nullifying the poisons of snake bites if this giant water bug's head either fresh or dried is immediately pasted on the bitten part. Miraculously the insect's head extracts the poisonous blood by oozing it out. (There is a legend about the story of the giant water bug and a snake, and it is according to this belief that the people used it).

Many tribal communities highly treasured the thunderbolt stones (*Braja nung)* which come in different shapes, sizes and colours. The metamorphic rock is rub against hard stone with little water and the extracted liquid is drank by adding little water and some liquid are also applied on certain body parts like forehead, chest and on the back areas in cases of high fever, sudden stomach ache when they believe it was due to the contact of evil spirits.

They also highly valued various animals' bile especially the bear, tiger and python's bile in the preparation of certain indigenous medicines. These dried animal's biles are used in small dosage especially, for measles and other related health issues when needed.

Mrs. K. Tomui (78/F) of Kharou Khunou said that "The meat of the wild boar (War hog) like creature was eaten in the past to cure the leprosy (*Laichi* in Maring). Even the fats were used to apply on the leprosy infected body parts. In doing so, the patient gets cured within a week's time", (see Table 5, Sl. No.6).

Besides, these few examples there are many materials and objects used by the Maring and other tribal communities of Northeast India in the application of treating different ailments and diseases. The nature of treatment methods varies from priest to priest and region to region and community to community.

The Table 12 is a paradigm that highlights how the Maring tribe used various compound plants often by adding certain essential items like honey, pure mustard oil or crystal cane sugar (*Sitamasi* or *Misri* in Assamese) in healing and curing various types of ailments, sicknesses and diseases. The essential items are used to enhance the taste or reduced the bitter or sour or clumsy taste without compromising the quality of the medicines. It may be pointed out that when the ailment, sickness or diseases had been prolonged for some time the Maring/ local medicine men preferred to use compound medicines since the composition is considered more powerful and effective in fighting against the dreaded sicknesses or diseases than single plant item. Often the dosages are lesser in quantity than the single plant dosages as observed.

Conclusion

The study is based on a survey to understand how extensive and relevant the Maring community used the ethno-botanical plants as medicines in the present perspectives since, many tribes and communities especially, of the Northeast India and rest of the world are declining in practice of their indigenous medicinal values system.

According to Mrs. Hoinu of Pallel a respondent said that "There are many Maring (and Khoibu) villages in the region but the number of well knowledgeable persons in the field of ethno-medicines have considerably declined in the recent years and it is feared that the remaining few people would remember such ethno-medicine if not recorded soon".

In the study, 95 ethno-medicinal plants are listed out of which 60 species are identified belong to 37 families while the 35 plants need further investigation. The most common ailment and diseases are like fever, cough, urinary/kidney stones case, irregular menstrual problems, blood purifier for post-natal woman, diabetes, cancer, jaundice, tonsillitis, chronic ulcer and so on. It is found that the bulk of the indigenous plants belong to herbs and shrubs which are seasonal in nature. Of all these ethnomedicinal plant they used maximum leaves part that is often taken in the traditional oral form of decoction and syrup. Further, these ethnomedicinal plants are considered harmless and effective while others considered deadly are used for external applications.

In fact, with the increased influences of modernization, westernization, globalization, modern education and Christianity many younger generation prefer to use pharmaceutical medicines like for headache, stomach-ache. But this does not mean that the people have completely given up their traditional indigenous medicines. Depending upon the degree and types of illness, in certain cases magico-religious ceremonies are also employed like the Nigerians of Africa along with the ethno-medicinal plants by the local medicine-man who also act as priest and priestess in the healing and curing various dreaded diseases.[2]

However, it is observed that no Maring individual has a herbarium at home apart from few plants planted in the courtyard or garden.

There is no doubt that the Maring (and Khoibu) people even today continued to use their indigenous local made medicines in the treatment of simple to complicated ailments and diseases besides, the availability of modern pharmaceutical medicines. In short, they used side by side both ethno-medicine and pharmaceutical medicines. The reason of continuity in their usage of these ethno-medicines is because they considered effective in curing and healing certain common sicknesses in some respect. It is also found that the Maring used more single plants in the treatment of common sicknesses and compound plants often in the treatment of complicated and serious ailments.

Besides, the listed ethno-botanical plants of Maring tribe there are still many more ethno-medicinal plants that need to be scientifically studied and identified including the constituent components and elements systematically. The above study is a survey to explore how intensive and extensive the Maring tribe used such ethno-botanical plants as medicines in their daily lives.[3]

References

1. Maring Naga. http://en.wikipedia.org/wiki/Maring_Naga, accessed on 31/12/12.
2. Grierson, George Abraham. *Linguistic Survey of India: Tibetan-Burman Family,* Vol. III, Part-III (Specimens of the Kuki-Chin and Burma groups). Calcutta: Superintended of Government Printing; 1967(reprinted, Delhi: Motilal Banarsidass).
3. Census of India, 2011.
4. Brown, R. *Statistical account of native states of Manipur and the hill territory under its rule.* 1873 Delhi: Sanskaran Prakashak.
5. Singh BH, PK Singh, SS Singh & Elangbam B. Indigenous bio-folklores and practices: Its role in biodiversity conservation in Manipur. *Journal of Hill Research.*1996, 9(2):359-362.

6. Ningombam DS, Deshworjit Singh & Potsangbam Kumar Singh. Ethno-botanical study of *Phologa-canthus thyrsiformis Nees*: A conserved medicinal plant of Manipur, Northeast India. *International Journal of Herbal Medicine*. 2014, 1(5): 10-14.
7. Yuhlung, Cheithou Charles & PG Jangamlung Richard. *Northeast India tribal studies: An insider's view*. New Delhi: Regency Publication and Astral Publication; 2015, ix.
8. Mao, AA; TM Hynniewta & M Sanjappa. Plant wealth in Northeast India with reference to ethnobotany. *Indian Journal of Traditional Knowledge*.2009, 8(1): 96-103.
9. Wangkhem Indira Devi1, Guruaribam Shantibala Devi & Chingakham Brajakisor Singh. Traditional herbal medicine used for the treatment of diabetes in Manipur, India. *Research Journal of Pharmaceutical, Biological and Chemical Sciences*. 2011, 2(4): 709-715.
10. Rajkumari, Ranjana; PK Singh, Ajit Kumar Das & BK Dutta. Ethno-botanical investigation of wild edible and medicinal plants used by the *Chiru*Tribe of Manipur, India. *Pleione*. 2013, 7(1): 167-174.
11. Sobita K & Th Bhagirath. Effects of some medicinal plant extracts on *Viciafaba* root tip chromosomes. *Caryologia*. 2005, 58(3): 255-261.
12. Yumnam JY & OP Tripathi. Traditional knowledge of eating raw plants by the Meitei of Manipur as medicine/ nutrient supplement in their diet. *Indian Journal of Traditional Knowledge*. 2012, 11(1): 45-50.
13. Rajesh Singh Yumnam, Chonita Devi, Santosh Kumar Singh Abujam & D Chetia. Study on the ethno-medicinal system of Manipur. *International Journal of Pharmaceutical & Biological Archives*. 2012, 3(3): 587-591.
14. Yumnam JY & OP Tripathi. Traditional knowledge of eating raw plants by the Meitei of Manipur as medicine/ nutrient supplement in their diet. *Indian Journal of Traditional Knowledge*. 2012, 11(1): 45-50.
15. Ridip Hazarika, Santosh Kumar Singh Abujam & Bijoy Neog. Ethno-medicinal studies of common plants of Assam and Manipur. *International Journal of Pharmaceutical & Biological Archives*. 2012, 3(4): 809-815.
16. Khongsai M, Saikia, SP & Kayang. Ethnomedicinal plants used by different tribes of Arunachal Pradesh. *Indian Journal of Traditional Knowledge*. 2011, 10(3): 541-546.
17. Jamir NS, Lanusunep & Narola Pongener. Medico-herbal medicine practiced by the *Naga* tribes in the state of Nagaland (India). *Indian Journal of Fundamental and Applied Life Sciences*.2012, 2(2): 328-333.

18. Jamir NS, Lanusunep & Narola Pongener. Medico-herbal medicine practiced by the *Naga* tribes in the state of Nagaland (India). *Indian Journal of Fundamental and Applied Life Sciences.* 2012, 2(2): 328-333.
19. Neli Lokho Pfoze, Yogendra Kumar & Bekington Myrboh. Survey and assessment of ethno-medicinal plants used in Senapati District of Manipur State, Northeast India. *Phytopharmacology.*2012, 2(2): 285-311.
20. Mao AA; TM Hynniewta & M. Sanjappa. Plant wealth in Northeast India with reference to Ethnobotany. *Indian Journal of Traditional Knowledge.*2009, 8(1): 96-103.
21. Ningombam DS, SP Devi, PK. Singh, A Pinokiyo & Bisheswori Thongam. Documentation and assessment on knowledge of ethno-medicinal practitioners: A case study on local Meetei healers of Manipur in Journal of Pharmacy and Biological Sciences (IOSR-JPBS). 2014, 9(1): 53-70.
22. Yuhlung, Cheithou Charles & Mini Bhattacharyya. Ethno-medicine practice among the Chothe tribe of Manipur, North-east India in Tribal Health Bulletin. 2015, 22 (1&2): 40-58.

Suggestion

1. The concern Departments and agencies of Central and State Governments should immediately address an urgent and serious survey to scientifically study and document systematically the ethnomedicinal plants especially, among the indigenous tribes of Northeast India being situated in one of the world's hotspot zones.
2. Through Government and institutions atleast one or two villages of a tribe especially, of the Northeast India needs funding to cultivate herbarium of ethno-medicinal plants as part of encouragement to communities in preservation of its values. The focus should be on more on hill tribes, since most of the endangered ethnomedicinal plants are grown in the deep hilly forest areas.
3. The Government should pass an Act to amend immediately for the protection and conservation of certain forest areas where identified endangered ethnomedicinal plants are grown in the deep interior forest. Since, certain ethno-botanical plants grows naturally only where an environment suits them for their growth and if cultivated it retards the growth or dies or did not produce an equal result.
4. Frequent workshops and seminars on awareness programmes should be conducted at the sensitive village zone in each district time to time by concern Departments of Central and State agencies to sensitize about the life saving plants and its values as part of conservation and preservation of ecology and environments, thereby protecting certain endangered flora and fauna too.
5. A separate website for only the Northeast India communities on ethnomedicines should be created by direct website and hyperlinks with other related subject matters especially, for scholars, students, and concern public for their future references.

****

Researcher with the Respondents of Maring (and Khoibu) Tribe, Manipur

****

www.ingramcontent.com/pod-product-compliance
Ingram Content Group UK Ltd.
Pitfield, Milton Keynes, MK11 3LW, UK
UKHW021011290726
14059UKWH00001BA/79

9 789390 384181